HANDBOOK OF MATERIALS TESTING REACTORS AND ASSOCIATED HOT LABORATORIES IN THE EUROPEAN COMMUNITY

HANDBOOK OF MATERIALS TESTING REACTORS AND ASSOCIATED HOT LABORATORIES IN THE EUROPEAN COMMUNITY

revised edition
February 1981

Abstract

Technical information is given on the large EEC's Materials Testing Reactors (thermal power > 5 MW) and associated hot cell facilities, with the main emphasis on experimental irradiation facilities, specialized irradiation devices (loops and capsules) and possibilities for post-irradiation examinations of samples. The information is presented for each reactor in the form of eight information sheets under the headings: main characteristics of the reactor; utilization and specialization of the reactor; experimental facilities of the reactor; neutron spectra; main characteristics of specialized irradiation devices; main characteristics of hot cell facilities; equipment and techniques available for post-irradiation examinations; utilization and specialization of the hot cell facilities.

Commission of the European Communities

HANDBOOK OF MATERIALS TESTING REACTORS AND ASSOCIATED HOT LABORATORIES IN THE EUROPEAN COMMUNITY

Nuclear Science and Technology

Edited by

PETER VON DER HARDT and HEINZ RÖTTGER

C.E.C., Joint Research Centre, Petten Establishment

D. REIDEL PUBLISHING COMPANY

DORDRECHT : HOLLAND / BOSTON : U.S.A.

LONDON : ENGLAND

Library of Congress Cataloging in Publication Data

DATA APPEARS ON SEPARATE CARD.

CIP

ISBN 90-277-1347-2

Publication arrangements by
Commission of the European Communities
Directorate-General Information Market and Innovation, Luxembourg

EUR 5369e (revised edition 1981)

Published by D. Reidel Publishing Company
P.O. Box 17, 3300 AA Dordrecht, Holland

Sold and distributed in the U.S.A. and Canada
by Kluwer Boston Inc.,
190 Old Derby Street, Hingham, MA 02043, U.S.A.

In all other countries, sold and distributed
by Kluwer Academic Publishers Group,
P.O. Box 322, 3300 AH Dordrecht, Holland

D. Reidel Publishing Company is a member of the Kluwer Group

Printed in The Netherlands

INTRODUCTION

The original "MTR Handbook" was compiled as a result of the work of the Coordination Committee on Materials Testing Reactors, in 1973/74. The work of the Committee has not been pursued beyond 1974, but the Handbook has nevertheless been updated March 1975 and edited.
Today, about six years later, another group has updated the information of the initial handbook, viz. the Irradiation Devices Working Group (IDWG). As in the past, the review has been limited to large research and test reactors with more than 5 MW of thermal output. Three new facilities which are not materials testing reactors have been included: the HFR Grenoble, the ORPHEE Saclay and the KNK II Karlsruhe. The first two are high power reactors for fundamental research and the third one is a fast reactor demonstration plant.

After the initial enthusiasm of the 1960's, research and test reactors have passed through critical years during the 1970's. Many of them have been shut down or considered for shut-down. The present decade started with a generally more positive situation: major modifications, upgrading, and modernisation have been carried out or are planned on reactors and their experimental facilities.
Today, the Countries of the European Communities display an impressive and unique set of large testing stations, equipped with the most advanced experimental installations, and capable of handling any type of work that can be done with neutrons: fundamental and technological materials research; radioisotope production for industrial, medical and scientific purposes; nuclear and non-nuclear application of neutron radiography; solid state and nuclear physics research; activation analysis, training and eduction; and others.

The editors wish to acknowledge the spontaneous and frank collaboration of their colleagues from research institutes in the Community which made this Handbook a valuable contribution to nuclear science and technology.

P. von der Hardt
H. Röttger

TABLE OF CONTENTS

*) *All information sheets have been revised in February 1981.*

1. THE BR 2 REACTOR AT MOL, BELGIUM AND ASSOCIATED HOT CELL FACILITIES. INFORMATION SHEETS.

Fig. 1.1 The BR 2 reactor at CEN/SCK Mol, Belgium.

1.1 INFORMATION SHEET No. 1. MAIN CHARACTERISTICS OF THE REACTOR.

Name of reactor	BR 2
Organization	Centre d'Etude de l'Energie Nucléaire (CEN) / Studiecentrum voor Kernenergie (SCK)
Site	MOL - CEN/SCK - Belgium
In operation since	1961
Power MW (th)	80 to 125
Moderator	H_2O + Be
Coolant	H_2O
Fluid and pressure in experiment channels	H_2O at 12 bar. The head of a rig is exposed only to the static pressure of the pool water, i.e. 0.6 bar.
Nominal operation (days/year)	220 - 250
Operating cycle (days)	
on power	21
shut down	7

Responsible officers (name, address):

BR 2	F. Léonard	CEN/SCK, 200 Boeretang, B-2400 Mol
Hot Lab ATHA	F. Léonard	
Hot Lab LHMA	A. Demildt	

1.2 INFORMATION SHEET No. 2 UTILIZATION AND SPECIALIZATION OF THE REACTOR

1.2.1 Original purposes of the reactor

- Materials testing (in the core)
- Isotope production (in the reflector)
- Fundamental physics research (in beam tubes)

1.2.2 Major modifications subsequently made

- The original alloy type fuel elements, containing 244 g ^{235}U, have been replaced by cermet elements, of identical geometry, containing 400 g ^{235}U plus burnable poisons.
- The original design power of 50 MW has been uprated to 125 MW by provision of new heat exchangers and by changing the cooling tower packing.
- Beryllium matrix replacement.

1.2.3 Major modifications planned

none

1.2.4 Major projects and programmes for which the reactor presently is used

- Sodium cooled fast reactor development ≈ 35 °/o
- Gas cooled fast reactor and high temperature gas reactor programmes ≈ 50 °/o
- Various irradiation programmes (water reactors, radioisotopes, etc.) ≈ 15 °/o

1.2.5 Loops and irradiation devices specially developed for the reactor

see chapter 1.5

1.2.6 Additional information

o special features of the reactor design

- the experimental channels are skew, the tube bundle presenting the form of a hyperboloid of revolution (see Fig. 1.2). This gives easy access at the top and bottom reactor covers allowing complex instrumented devices, while maintaining a very high neutron flux at the core.
- great flexibility of utilization, due to the fact that it is possible to adapt the core configuration to the experimental loading as the fissile charge can be centred on different experimental channels.
- although BR 2 is a thermal reactor, it is possible to achieve neutron spectra very similar to those obtained in a fast reactor, either by the use of absorbing screens or by the use of fissile material within the experimental devices (converters).
- five 200mm diameter channels are available for loading large experimental irradiation devices, as in-pile sodium, gas or water loops.

o present purposes of the irradiations

- study of the behaviour of fuel elements and structural materials intended for the reactors of future nuclear power stations (sodium or gas cooled fast reactors, high temperature gas cooled reactors, light water reactors, fusion reactors).
- safety in-pile experiments (particularly restriction to fuel pin cooling and transient over-power) for these reactors.

- *production of high specific activity radioisotopes.*
- *basic physical research within the beam tubes.*
- *neutron radiography in the reactor pool.*
- *gamma irradiations within spent fuel elements.*

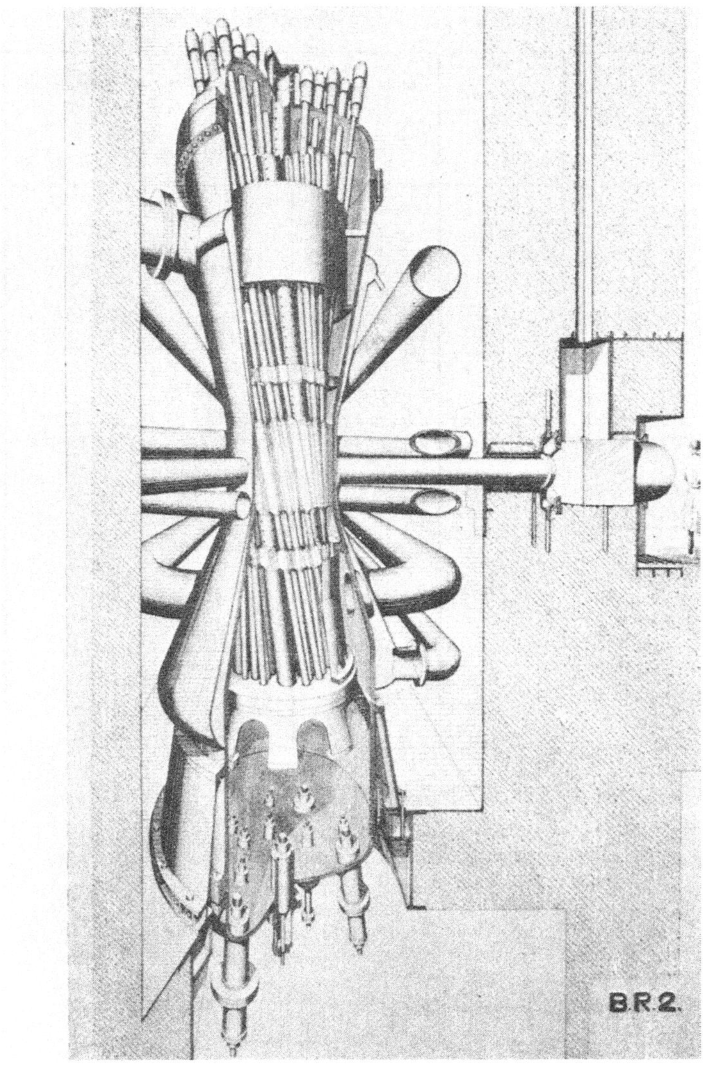

Fig. 1.2 General view of the BR 2 reactor.

1.3 INFORMATION SHEET No. 3. EXPERIMENTAL FACILITIES OF THE REACTOR.

1.3.1 Irradiation Positions

A summary of the data concerning the irradiation positions is given in the following table :

Table 1.1 Irradiation positions for core configuration 9 B or 9 C

type/name	number (1)	useful diameter mm	typical neutron flux density in $m^{-2}s^{-1}$ (2)		nuclear heating W/g Al
			thermal (3)	fast (E>0.1 MeV)	
in the pressure vessel					
o core					
- within the fuel elements	37	17,4 to 42,4	$3,8 \times 10^{18}$	7×10^{18}	15
- in the 200 mm channels	1 or 2	up to 200	-- (4)	7×10^{18}	13
o reflector					
- 200 mm holes	up to 4	29,5 to 200,2	9×10^{18} (5)	$1,5 \times 10^{18}$	6
- 85 mm holes	19 to 31 (6)	29,5 to 80,6	5×10^{18}	2×10^{18}	6
- 50 mm holes	10	29,5 to 46	$1,6 \times 10^{18}$	$0,6 \times 10^{18}$	2
outside the pressure vessel					
o beam tubes	9	up to 305	10^{11} to 5×10^{17}	--	--
o reactor pool					
- radioisotopes tubes	4	80	3×10^{17}	5×10^{16}	< 0,5
- neutron radiography	1	130	3×10^{11}	-	--
- gamma facility	5	12 to 80	--	--	5×10^{7} rad/h

Remarks :
(1) total number of irradiation positions: up to 100
(2) 84,5 MW reactor power. 106 MW in core configuration 9 C with a driver fuel element in channel H1
(3) unperturbed, measured at the axis, $[n]_0^{0,5}$ at 2 200 m/s
(4) under cadmium screen
(5) in the central channel loaded with beryllium
(6) one 200 mm channel is equivalent to three 85 mm channels.

1.3.2	Reactivity available for experiments:	up to 7 % $\Delta k/k$
1.3.3	Neutron spectra (data):	see chapter 1.4
1.3.4	Specialized equipment available for experimenters:	- Neutron radiography of active or inactive rigs - Underwater Television camera
1.3.5	Availability of facilities for irradiation on behalf of other organizations of the European Communities:	Facilities are available

1.3.6 *Availability of hot cell facilities:* *Hot cell facilities are available on reactor site, see chapter 1.6.*

1.3.7 *Relevant publications:* *see chapter 1.5.*

1.4 *INFORMATION SHEET No. 4. NEUTRON SPECTRA (DATA)*

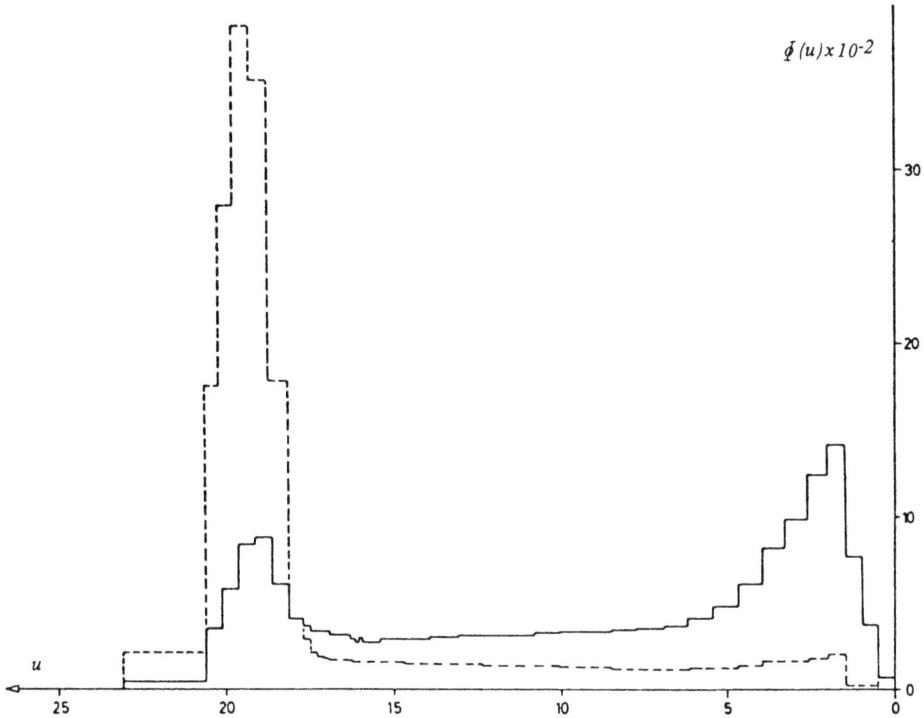

Fig. 1.3 *$\Phi(u) \times 10^{-2}$ neutron spectrum in BR2:*
—— along the axis of a fresh fuel element containing 400g U235.
----- along the axis of a reflector channel adjacent to the core.

Table 1.2: Typical neutron spectra for the BR 2 reactor

u (upper limit)	Energy (lower limit)	ϕ (u) . 10^{-2} centre of fresh cermet element 400g U235	ϕ (u) . 10^{-2} Be reflector channel
23,106	0,001 eV	0,434	2,132
20,580	0,0125	3,546	17,448
20,080	0,0206	5,782	27,968
19,580	0,034	8,312	38,182
19,080	0,0560	8,666	35,118
18,580	0,0924	6,082	17,850
18,080	0,152	4,145	6,055
17,679	0,227	3,665	2,911
17,476	0,278	3,388	2,129
17,275	0,340	3,284	1,898
17,078	0,414	3,291	1,847
16,889	0,50	3,091	1,708
16,250	0,9475	2,779	1,572
16,206	0,99	2,929	1,465
16,166	1,03	2,764	1,519
16,119	1,08	2,675	1,433
16,056	1,15	2,959	1,626
15,933	1,30	2,765	1,453
15,43	2,15	2,868	1,500
14,66	4,65	2,900	1,470
13,89	10	2,975	1,443
13,12	21,5	3,062	1,417
12,35	46,5	3,114	1,382
11,58	100	3,175	1,349
10,81	215	3,258	1,319
10,04	465 eV	3,288	1,279
9,27	1,00 keV	3,340	1,243
8,50	2,15	3,418	1,209
7,73	4,65	3,473	1,173
6,96	10	3,651	1,166
6,19	21,5	4,106	1,187
5,42	46,5	4,804	1,236
4,65	100	6,094	1,374
3,96	200	8,306	1,630
3,27	400	9,796	1,648
2,58	800 keV	12,405	1,791
2,01	1,40 MeV	14,137	1,998
1,44	2,50	7,694	1,342
0,96	4,00	3,717	0,271
0,48	6,50 MeV	0,696	0,058

1.5 INFORMATION SHEET No. 5. MAIN CHARACTERISTICS OF SPECIALIZED IRRADIATION DEVICES

1.5.1 Fuel-pin irradiation

o sodium loops with cadmium screen

MFBS:	up to 250 kW, up to 7 pins (oxide 600 W/cm, carbide 1200 W/cm)
MOL 7B:	up to 500 kW, 19 pins (oxide 600 W/cm) or 7 pins (carbide 1500 W/cm)
MOL 7C:	up to 700 kW, 37 pins bundle, safety experiment (Na flow failure test)

o helium loop with cadmium screen

GSB:	400 kW, 61 atm, 12 vented pins

o water loop

WRS (in design) :	LOCA and ATWS test on 45 pins bundle

o rig (1 pin)

FAFNIR and FASOLD capsules: (with Cd screen)	oxide, 6mm diameter, 350 to 600 W/cm, or carbide, 8 mm diameter, 1000 to 1500 W/cm fission product pressure measurement, central temperature measurement.
CIRCE and CEB (thermal flux):	continuous measurement of the power, possibility of detecting cladding rupture.
CFC:	compatibility fuel canning test.
THEBE:	thermal conductivity measurement.
P. basket:	test of advanced type LWR fuel rod
PWC :	pressurized water capsule (up to 150 kg/cm^2).
VNS device:	^3He screen for power cycling.
VIC:	compact sodium loop designed to perform different kinds of transient operating conditions
GBII:	vented pin irradiation under cadmium.
FARFADET:	start-up conditions tests on LMFBR fuel rod (under cadmium).

1.5.2 Fissile material irradiation

o CPUR and MOPS:	swept loops for irradiation of coated particles.
o POM:	high burn-up (100 000 MWd/t) in short time (a few months) with continuous measurement of the power.
o MOL 12:	creep experiment with in-pile measurement.
o MOL 13:	hollow fuel pellet rig (oxide, 1000 W/cm).
o MOL 10 and CIRCES:	in-pile swelling measurement.

1.5.3 Cladding materials and structural material irradiation

- *o MOL 1:* *samples directly cooled by the reactor water (<100°C)*
- *o MOL 3 and DISP:* *in NaK or Na, 600 to 700°C, under cadmium in several cases.*
- *o MOL 2 and 5:* *creep experiment with in-pile measurement.*
- *o MGR:* *graphite irradiation, 150 to 1200°C.*
- *o IPCTL:* *helium loop for graphite irradiation (carbon transfer experiment).*

1.5.4 Radioisotope production

- *o PACT:* *production of transplutonium elements*
- *o CSF:* *standard capsule*
- *o Actinium thimble:* *irradiation of $RaCO_3$ with double containment*
- *o Co and Ir capsules*
- *o Hydraulic rabbit; self-service thimble and pool tubes.*

1.5.5 Relevant Publications

o Irradiation devices

- *J. Planquart et al.*
  In-pile experimental equipment for BR2.
  Third U.N. International conference on the peaceful uses of atomic energy, Geneva, 1964.
- *J. Planquart*
  Dispositifs d'irradiation pour BR2.
  EAES Symposium on irradiation experiments, The Hague, 1965.
- *M. Soenen*
  Experience gained with the in-pile sodium loop for fast neutron irradiation after the first run in the BR2 reactor.
  Symposium on alkali metal coolants corrosion studies and system operating experience, Vienna, 28 Nov. - 2 Dec. 1966.
- *J.M. Baugnet*
  Boucles d'irradiation au réacteur BR2.
  HFR Information Meeting, Petten, 7 - 8 Dec. 1967.
- *P. von der Hardt*
  Development and in-pile performance of some BR2 irradiation rigs.
  EUR 3626e, 1967.
- *W. Hebel*
  Production d'éléments lourds par irradiation d'américium 241 dans le réacteur BR2.
  EUR 3560f, 1967.
- *M. Soenen*
  Experience gained with the 200 kW integrated in-pile sodium loop in the BR2 reactor.
  International conference on fast reactor irradiation testing, Thurso, 14 - 17 Apr. 1969.
- *G. Vanmassenhove*
  In-pile sodium loop 500 kW. Conceptional design.
  International conference on fast reactor irradiation testing, Thurso, 14 - 17 Apr. 1969.

- *P. von der Hardt*
  BR2 irradiation of fuel pins in liquid metal rigs.
  International conference on fast reactor irradiation testing, Thurso, 14 - 17 Apr. 1969.
- *A. Lhost*
  Thermal neutron fuel irradiation rig with continuous heat flux measurements.
  International conference on fast reactor irradiation testing, Thurso, 14 - 17 Apr. 1969.
- *J. Debrue, Ch. De Raedt, H. Lenders, F. Léonard, N. Maene, F. Motte, G. Stiennon.*
  Utilization of the BR2 materials testing thermal reactor for fast neutron irradiation.
  Fourth U.N. international conference on the peaceful uses of atomic energy, Geneva, 6 - 16 Sept. 1971.
- *P. von der Hardt*
  Some BR2 irradiation devices for fast reactor fuel and fuel elements.
  EUR 4632e, 1971.
- *P. von der Hardt*
  BR2 irradiation devices for HTGR fuel
  EUR 4737e, 1971.
- *W. Hebel, J. Planquart, G. Vanmassenhove*
  Irradiation devices for BR2 reactor - Design and operating characteristics.
  EUR 5543e MF (1976)
- *G. Vanmassenhove, A. Verwimp, W. Kramer*
  An in-pile sodium loop for investigation of local cooling disturbances in LMFBR's.
  25th meeting Euratom irradiation devices working group, Petten 18 - 19 May 1978.

o *Reactor features and programmes*

- *K.J. Bobin, H.J. Lenders*
  Improved performance fuel elements for BR2
  Nuclear Engineering Intern. 18, (1973), p. 418 - 421.
- *J. Debrue, Ch. De Raedt, N. Maene*
  Prediction of the irradiation characteristics of fast reactor fuel pins tested in cadmium screened loops in BR2.
  Groupe de travail Euratom sur les dispositifs d'irradiation, Saclay, 24 - 25 May 1973.
- *P. van Assche (ed.)*
  Research with BR2 Neutron Beams.
  BLG 519 (1977)
- *CEN/SCK*
  BR2 safety reassessment.
  Febr. 1977.
- *J.M. Baugnet, J.M. Gandolfo, W. Boeykens*
  The BR2 reactor and its associated facilities.
  Annual progress report 1977.
- *J.M. Baugnet, F. Léonard, J.M. Gandolfo*
  Bilan de 15 années d'utilisation du réacteur BR2 et programme futur.
  Symposium sur l'avenir des réacteurs de recherche, Grenoble 16 - 17 Nov. 1977.
- *J.M. Baugnet*
  Main features of the BR2 reactor and, in particular, of its fuel cycle.
  IAEA Advisory group meeting on research reactor renewal and upgrading programmes, Vienna, 23 - 26 May 1978.
- *CEN/SCK, KfK, KFA*
  Proceedings of the BR2 reactor review meeting.
  MOL, June 1, 1978.
- *J.M. Baugnet, W. Boeykens, J.M. Gandolfo, H. Lenders, F. Léonard*
  General outline of the operation and utilization of the BR2 reactor. Present situation and future trends.
  71/JMB/RH/172/80e, June 1980.

1.6 *INFORMATION SHEET No. 6. MAIN CHARACTERISTICS OF HOT CELL FACILITIES*

Table 1.3: Main characteristics of hot cell facilities

name of laboratory	year of active oper.	scope		gamma shielding (material, thickness, density)	leak tightness	special atmosphere	manipulators no., type (incl. spare manipul.)
		active area (m^2) (1)	number of working positions				
Laboratory for High and Medium Activity L.H.M.A.	1965	203	32 + 9 under constr.	new cells 1 m concrete (barytes) 3.5 g/cm^3 old cells, lead 0.1 to 0.2 m	all facilities treating U or Pu are α-tight only those facilities treating solely reactor constr. materials are βγ shielded without α-tightness	most facilities can be used under N_2 (100 ppm) dried air (100 ppm) or natural atmosphere (50 to 100 % humidity	1 - MG 12 - M7 30 - MA 11 6 - BF 700 4 - M8 4 - ME
Atelier Très Haute Activité A.T.H.A.	1965	48	5	concrete 1.37 m 3.4 g/cm^3	not α-tight (2)	no special atmosphere	11 - M8 6 - M9 1 - GM 300

Remarks: (1) inside hot cells
(2) special equipments are available for work on α-contaminated pieces.

1.6.1 Owner/Operator

L.H.M.A. building : CEN/SCK - Division of Applied Research

A.T.H.A. building : CEN/SCK - BR2 Department

1.6.2 Site

Mol, CEN/SCK, Belgium

1.6.3 Associated reactor

L.H.M.A.: not especially associated with a reactor

A.T.H.A.: especially associated with BR2.

1.7 *INFORMATION SHEET No. 7. EQUIPMENT AND TECHNIQUES AVAILABLE FOR POST-IRRADIATION EXAMINATIONS*

L.H.M.A.

- *Monocular and stereo visual inspection and photography*
- *Mechanical or chemical sampling of chemical precipitations upon the fuel pins.*
- *Chemical analysis of precipitations upon the fuel pins.*
- *Length measurement, profilometry and eddy current measurements upon fuel pins.*
  N.B.: very soon gamma activity measurement will be possible.
- *Fission gas sampling, analysis and internal volume measurements of fuel pins.*
- *Marking and cutting of longitudinal and radial samples of fuel pins.*
- *Metallography of fuel and cladding, as well as micro-hardness measurements.*
- *Chemical analysis by microprobe and scanning electron-microscope equipped with an X-ray fluorescence instrument.*
- *Chemical thinning and preparation of samples for transmission electron microscopic examination.*
- *Post-irradiation thermal treatment.*
- *Preparation of samples and performing of tensile strength, burst and charpy measurements.*
- *Axial and bi-axial creep tests and burst test measurements.*
- *Burn-up analysis (in cooperation with the Chemistry Department)*
- *Cladding and fuel density determinations.*
- *Analysis of hydrogen, tritium and He in cladding material and fuel.*
- *Analysis of iron and manganese 54 in cladding material.*
- *Making and application of protective layers with epoxy material.*
- *Dimensional measurements of grids.*
- *Measurement of the elasticity of the springs in the grids.*
- *Chemical kinetic studies with a hot thermal balance facility.*
- *Chemical preparative work upon plutonium and α,β,γ-auto-radiographies of irradiated fuel are performed in 7 glove-boxes.*

A.T.H.A.

- *Monocular and stereo visual inspection and photography.*
- *Length measurement and profilometry upon fuel pins.*
- *Gamma scanning.*
- *Preparation of neutron radiography*
- *Dry sipping*
- *Puncture and fission gas sampling.*
- *Dismantling of fuel assemblies and non-destructive tests.*
- *Dismantling of rigs and loops containing sodium or NaK (up to 10l.)*
- *Dismantling of aluminium capsules containing cobalt grains or discs, ^{60}Co measurement.*
- *Maintenance (including loading of new sources) of gamma-cells.*
- *Preparative work for long term storage or waste disposal.*
- *Loading of irradiated fuel pins into irradiation devices.*

1.8 INFORMATION SHEET No. 8. UTILIZATION AND SPECIALIZATION OF THE HOT CELL FACILITIES

1.8.1 Original purposes of the facilities

Post-irradiation examination of several materials.

L.H.M.A. building - non-destructive and destructive tests.

The LHMA can treat samples from fuel pins of thermal reactors, pressurized water reactors, boiling water reactors and fast reactors. The oxide or carbide fuel can be treated under inert atmosphere if necessary.

A.T.H.A. building - essentially non-destructive tests and dismantling operations.

1.8.2 Major modifications subsequently made

Adaptations to new techniques and fuels; for example carbide fuels and H.T.R. fuels.

1.8.3 Major modifications planned

L.H.M.A. - Extension of hot facilities for non-destructive and destructive testing of fuel pins for power reactors (thermal and fast).

A.T.H.A. - Extension of hot facilities for non-destructive testing and dismantling of fuel assemblies for power reactors.

1.8.4 Major programmes for which the facilities presently are used

- *Fast breeder reactor fuel pin development programmes (oxide and carbide of U and Pu).*
- *High temperature reactor fuels (coated particles).*
- *Thermal and power reactor fuel pins.*
- *Examination of structural materials (candidate cladding materials, structural elements of fuel assemblies, reactor vessel surveillance programmes).*

1.8.5 Availability of the facilities for post-irradiation examinations on behalf of other organizations within the European Community.

The hot cell facilities are available.

2. THE DR 3 REACTOR AT RISØ, DENMARK AND ITS ASSOCIATED HOT CELL FACILITIES. INFORMATION SHEETS.

Fig. 2.1 The DR 3 reactor at the Risø National Laboratory.

2.1 INFORMATION SHEET No. 1. MAIN CHARACTERISTICS OF THE REACTOR.

Name of reactor	*DR 3*
Organization	*Risø National Laboratory*
Site	*Risø, Denmark*
In operation since	*1960*
Power MW (th)	*10*
Moderator	D_2O
Coolant	D_2O
Fluid and pressure in experiment channels	*Depending on experiment*
Nominal operation (days/year)	*305*
Operating cycle (days)	
on power	*23.5*
shut down	*4.5*

Fig. 2.2 DR 3: View into the reactor hall.

Responsible officers (name, address):

DR 3	- *H. Floto, Reactor DR 3*	*Risø National Laboratory, P.O. Box 49, DK - 4000, Roskilde*
Risø Hot Cell	- *H. Hougaard, Metallurgy Dept.*	

2.2 *INFORMATION SHEET No. 2 UTILIZATION AND SPECIALIZATION OF THE REACTOR.*

2.2.1 *Original purposes of the reactor*

Materials testing and isotope production

2.2.2 *Major modifications subsequently made with indication of the research programme concerned*

- *Beam experiments for solid state physics research governed by Risø National Laboratory have been installed at each end of the 4 horizontal, tangential experiment tubes (17.5cm Ø)*
- *A cold neutron source is installed in one of these tubes (see Fig. 2.3)*

Fig. 2.3 DR 3. Cold neutron source.

2.2.3 *Major modifications planned*

None

2.2.4 *Major projects and programmes for which the reactor presently is used*

- *Isotope production and industrial irradiation*
- *Fuel pin testing*
- *Basic research on solid state physics*

2.2.5 *Loops and irradiation devices specially developed for the reactor*

- *High pressure Rig HP 1 for fuel pin testing*
- *High pressure Rig HP 2 for fuel pin testing (also with thermal cycling)*
- *Corrosion Rig*
- *HTGR Rig for fuel irradiations up to 2000°C*

Main characteristics of specialized irradiation devices are given in chapter 2.5

2.3 INFORMATION SHEET No. 3. EXPERIMENTAL FACILITIES OF THE REACTOR

2.3.1 *Irradiation positions*

Table 2.1 : Irradiation positions

type/name	number	active length mm	useful diameter mm	typical unperturbed neutron flux density in $10^{18} m^{-2} s^{-1}$ thermal	fast ($E > 0.1$ MeV)	nuclear heating W/g Al
- fuel element central holes	26	610	50	1,5	0,45	1,5
7 V holes	4	450	175	0,6	0,005	0,3
4 V holes	4	150	100	0,3	0,001	0,2
7 T holes	4	135	175	0,7	0,002	0,1

2.3.2 *Reactivity available for experiments:* ~ 8% $\Delta k/k$

2.3.3 *Neutron spectra (data):* Not measured, see chapter 18.4 (PLUTO data)

2.3.4 *Specialized equipment available for experimenters:* Double beam neutron radiography facility at 2 kW reactor DR 1 at Risø

2.3.5 *Availability of facilities for irradiation on behalf of other organizations of the European Communities:* Facilities are available

2.3.6	*Availability of hot cells facilities:*	*Hot cells are available on the reactor site (see chapter 2.6)*
2.3.7	*Relevant publications:*	*Not communicated*

2.4 INFORMATION SHEET No. 4. NEUTRON SPECTRA (DATA)

Not measured, see for reference PLUTO data in chapter 18.4

2.5 INFORMATION SHEET No. 5. MAIN CHARACTERISTICS OF SPECIALIZED IRRADIATION DEVICES

2.5.1 Fuel pin irradiations

o High pressure Rig type HP 1	*1 pin, 200 W/cm^2, 9 - 15 mm Ø, length 600 mm, 150 bars pressurized water, failed fuel pins may be tested.*
o High pressure Rig type HP 2	*1 pin, 9 - 15 mm Ø, length 400 mm, 80 bars pressurized water, thermal cycling tests may be performed.*
o Corrosion Test Facility	*cladding samples 15 mm Ø, 500 mm length, 80 bars, 20 - 400°C water*
o HTGR fuel irradiation rig	*in fuel position, 700 - 2000°C*
o Structural materials irradiation rig	*reloadable, 40 - 500°C*
o Thermal conductivity rig	*for UO_2 up to 1900°C*

2.5.2 Relevant publications

- *H. Hagen, K. Hansen, J.A. Leth*
  Design and experience with rigs simulating LWR conditions in a research reactor (HP-rigs)
  IAEA - SM - 165/8, 1972.

2.6 INFORMATION SHEET No. 6. MAIN CHARACTERISTICS OF HOT CELL FACILITIES

Table 2.2 : Main characteristics of hot cell facilities

Name of Laboratory		Hot-Cell Facility
Year of active operation		1964
Scope	Number of working positions	7
	Active area	Caves and frogmens area : 432 m^2 (cave floor area 62 m^2) Room for glove-boxes and shielded cells : 334 m^2 Other blue and red areas : 2 634 m^2 Total, classified areas : 3 130 m^2
Gamma shielding-caves (material, thickness, and density)		Floor: Normal concrete, 0.8 m, 2.2 g/cm^3 Roof: Normal concrete, 1.0 m, 2.2 g/cm^3 Back and front wall: Normal concrete, 1,7 ~ 2.0 m, 2.2 g/cm^3 Walls between caves: Magnetite concrete, 1.0 m, 3.6 g/cm^3
Leak tightness		Yes - provided that all master-slave boots are intact.
Special atmosphere		Theoretically : yes In practice : no
Manipulators number and type		7 pairs of master-slaves, Nuclear Equipment heavy-duty, Harwell type α-seal. 1 power-operated manipulator, GEC Mk. 1 1 power-operated cell hoist, GEC, capacity : 1,5 t

2.6.1 Owner / operator — Risø National Laboratory, Metallurgy Department

2.6.2 Site — Risø, Risø National Laboratory

2.6.3 Associated reactor — DR 3 at Risø

2.7 INFORMATION SHEET No. 7. EQUIPMENT AND TECHNIQUES AVAILABLE FOR POST-IRRADIATION EXAMINATIONS

- *Maximum weight of transport containers is 25 t.*
  Max. length of fuel elements to be received is 5 m.
  Max. activity is 10^6 Ci (1 MeV) per concrete shielded cell, simultaneously.
- *Dismantling: Cutting machines, saws, drilling machines, tools, etc. are available*
- *Visual inspection: Monoculars, binoculars, periscope, techniscope, epitechniscope, polaroid cameras, highly sensitive TV-camera.*
- *Dimensional measurements: Profilometer (continuous of stepped -10 micron- diameter digital recording on a spiral), length measuring bench, Max. length 3.0 m.*
- *γ-scanning; γ-densitometry and γ-spectrometry: Stepped (10 micron) digital recording of single isotopes and relative density. Max. length 3.0 m. Use of multichannel analyzer and Ge-Li-detectors.*
- *Leak-testing: Bubble test, SF_6-test.*
- *X-ray radiography: available, but in practice superceded by neutron radiography.*
  High energy X-ray radiography for water-reactor fuel rods; low energy x-ray radiography for coated particle fuel.
- *Neutron radiography: for fuel pins up to 4.4 m.*
- *Fission gas sampling: Pressure and amount of fission gases measured; gas analysis by mass-spectrometry and gas chromatography. Residual gas in fuel samples is determined also.*
- *Metallography: Impregnation of fuel rods if necessary; grinding, polishing, etching; Reichert Telatom microscope.*
  Complete isotopic analysis by mass spectrometry and α-spectrometry (all U-, Pu-, Am- and Cm- isotopes).
- *Miscellaneous: Surface replication, extraction, replication and analysis by X-ray-diffraction or chemical methods; burn-up determination by Nd-method; H_2/D_2-determination by mass-spectrometry (Zr-alloys); autoradiography, α- and β/γ, mechanical testing of metals (steel samples, cladding tubes).*

2.8 INFORMATION SHEET No. 8. UTILIZATION AND SPECIALIZATION OF THE HOT CELL FACILITIES

2.8.1 Original purposes of the facilities

General purpose facility for examination of irradiated materials, for encapsulation of irradiation sources, and for repair work on contaminated/radioactive components.

2.8.2 Major modifications subsequently made

Six α-tight, lead shielded, stainless steel boxes established for isotope analyses, etc.

2.8.3 Major modifications planned

Only development and improvement of equipment are foreseen.

2.8.4 *Major programmes for which the facilities presently are used*

The Danish programme for studies and development on contract-basis of water-reactor fuel. Post-irradiation examination on contract-basis for foreign institutions.

2.8.5 *Availability of the facilities for post-irradiation examination on behalf of other organizations within the European Communities*

The facilities are available. During several years, more than 50 °/o of the work has been carried out on contract-basis for foreign firms and institutions.

2.8.5 *Relevant publications*

- *F. Bundgaard Larsen and Arne Sørensen.*
  Facilities at Research Establishment Risø for Post-Irradiation Examination of Dragon Project Experiments.
  Dragon Project Report No. 446, July 1966.

- *Danish Atomic Energy Commission, Research Establishment Risø.*
  Post-Irradiation Services at Risø.
  July 1969.

Fig. 2.4 Risø National Laboratory. Hot cells.

3. THE FR 2 REACTOR AT KARLSRUHE, F.R. GERMANY AND ASSOCIATED HOT CELL FACILITIES. INFORMATION SHEETS.

Fig. 3.1 The FR 2 reactor building (in the middle), at the site of KfK Karlsruhe.

3.1 INFORMATION SHEET No. 1. MAIN CHARACTERISTICS OF THE REACTOR.

Name of reactor	*FR 2*
Organization	*Kernforschungszentrum Karlsruhe GmbH*
Site	*Kernforschungszentrum Karlsruhe (KfK)*
In operation since	*1961*
Power MW (th)	*44*
Moderator	D_2O
Coolant	D_2O
Fluid and pressure in experiment channels	*Air, ~ 1 bar*
Nominal operation (days/year)	*220*
Operating cycle (days)	
on power	*40 days per operating cycle*
shut down	*14 - 49 (according to annual operation schedule)*

Responsible officer (name, address):

FR 2 *Hot Cells*	*D.I. Steiger*	*Kernforschungszentrum Karlsruhe GmbH* *Hauptabt. KTB* *Postfach 3640* *D - 7500 Karlsruhe*

3.2 INFORMATION SHEET No. 2 UTILIZATION AND SPECIALIZATION OF THE REACTOR

3.2.1 Original purposes of the reactor

- *Pure research*
- *Solid state physics,*
- *Materials research*
- *Isotope production*

3.2.2 Major modifications subsequently made

- *Reactor power increased from 12 MW to 44 MW*
- *Neutron flux increased from* $\phi_{th} = 3.10^{13}$ *to* 1.10^{14} $cm^{-2}s^{-1}$

3.2.3 Major modifications planned

- *Final shut-down in 1981*

3.2.4 Major projects and programmes for which the reactor presently is used

- *"Nuclear safety" project (PNS)*
- *"Fast breeder" project (PSB)*
- *Solid-state physics*
- *Materials research*
- *Isotope production*

3.2.5 Loops and irradiation devices specially developed for the reactor

- *Low-temperature irradiation facility (solid-state physics)*
- *Superheated steam circuit (transient burst experiments within reactor safety project)*
- *Various capsule tests (fuel element development)*
- *Pneumatic rabbit system(isotope production)*
- *Rotating irradiation devices in thermal column*

Main characteristics of specialized irradiation devices are given in chapter 3.5

3.2.6 Additional information

- *The FR 2 is the only materials testing reactor in the Federal Republic of Germany, which is operated with low enriched uranium fuel* (UO_2 *with 2°/o U-235).*
  This is regarded as an advantage with respect to both fuel supply and fuel reprocessing.
- *The large fuel zone (diameter ≈ 2.5m, height 2m) results in a flat radial and vertical distribution of the neutron fluxes, which is of special interest for irradiation of samples with larger dimensions.*

Fig. 3.2 FR 2 reactor. View into the reactor hall.

3.3 *INFORMATION SHEET No. 3. EXPERIMENTAL FACILITIES OF THE REACTOR.*

3.3.1 *Irradiation positions*

Table 3.1 : Irradiation position data

Type / name	*Number*	*Active length mm*	*Useful diameter mm*	*Typical unperturbed neutron flux density in $m^{-2}s^{-1}$* thermal	fast (E > 0.1 MeV)	*Nuclear heating W/g Al*
Isotope positions	41	2100	26	$5 - 9.5 \cdot 10^{17}$	$2 - 6 \cdot 10^{16}$	≤ 0.5
Free fuel element positions	20	2100	56	$4 - 9.5 \cdot 10^{17}$	$2 - 6 \cdot 10^{16}$	≤ 0.5
Thermal columns, vertical	6	1000	200x200 *)	$0.3 - 2.2 \cdot 10^{15}$		
horizontal	2	2000	100x100 *)			
	1	2000	600x400 *)			
	2	2000	200x400 *)			
	2	1000	200x200 *)			
Beamhole channels	2	200	290 / 330			
	7	300	130 / 190	$3 - 5 \cdot 10^{17}$	$1 - 2.5 \cdot 10^{16}$	0.1
	2	3000	55	$\leq 9 \cdot 10^{17}$	$\leq 6 \cdot 10^{16}$	≤ 0.5
	1	2000	55	$\leq 6 \cdot 10^{17}$	$\leq 4 \cdot 10^{16}$	≤ 0.3
Vertical channels	3	2100	130	$\leq 9 \cdot 10^{17}$	$\leq 6 \cdot 10^{16}$	≤ 0.5

*) *rectangular cross-section*

3.3.2 *Reactivity available for experiments:* ca. 5 °/o $\Delta k/k$

3.3.3 *Neutron spectra (data)* see chapter 3.4

3.3.4 *Specialized equipment available for experimenters:*

- *Neutron radiography,*
- *gamma scanning,*
- *gamma irradiation facility,*
- *hot cell in the reactor building,*
- *various transport casks,*
- *low-temperature transport cryostat,*
- *cold neutron source,*
- *facility for changing samples during operation,*

etc.

3.3.5 *Availability of facilities for irradiations on behalf of other organizations of the European Community:* *Facilities are available*

3.3.6 *Facilities available on the reactor site (see chapter 3.6):*

- *Hot cells of the KfK*
- *Transplutonium laboratory (TPL) of the KfK*

3.3.7 *Relevant publications:*

- *H. Oehme, W. Marth*
  Utllization of the reactor FR 2 as a research facility.
  Kerntechnik 8 (1966), no. 6, p. 253
- *KfK reports No. 688, 742, 746, 1415, 1705, 1956, 2071, 2154, 15/77-1, 15/78-1, 2821 B, 3032 B (Annual operating reports).*

3.4 INFORMATION SHEET No. 4. NEUTRON SPECTRA (DATA)

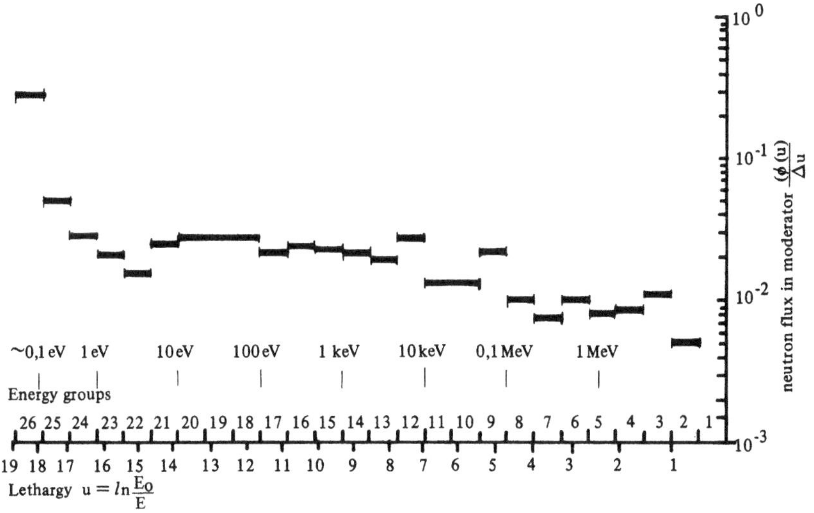

Fig. 3.3 Neutron energy spectrum of the FR2.

3.5 INFORMATION SHEET No. 5. MAIN CHARACTERISTICS OF SPECIALIZED IRRADIATION DEVICES

3.5.1 *Low-temperature irradiation facility*

Medium :	*He in gaseous form*
Expansion :	*reciprocating-piston device*
ϕ_{th} *(unperturbed):*	$8 \cdot 10^{17}\ m^{-2}s^{-1}$
ϕ_f:	$2.5 \cdot 10^{16}\ m^{-2}s^{-1}$
Refrigerating capacity at 12 K:	*500 W*
Lowest attainable irradiation-sample temperature:	*10 - 20 K according to reactor power. Facility for cold transport at 20 K*
Pressure experimental container:	*3,5 bar (abs)*
Publication:	*KfK Report No. 463*

Table 3.2 : Irradiation facilities

Type /names	Number	Useful dimensions	Nuclear Characteristics			Environmental Characteristics			Remarks
			Thermal neutron flux in $m^{-2}s^{-1}$	Fast neutron flux in $m^{-2}s^{-1}$ ($E > 0,1$ MeV)	Gamma flux MeV $cm^{-2}s^{-1}$	Max. allowed temp.	Max. allowed press.	Surr. medium	
<u>Facilities for routine irradiation</u>									
- air-cooled isotope production channels standard type	9	standard sample (modified harwell can): Φ 2,6 cm; l. 6,8 cm (longer samples possible) 29 standard samples per channel	$5 - 9,5 \cdot 10^{17}$	$2 - 6 \cdot 10^{16}$	$\sim 6,7 \cdot 10^{13}$	200^oC	ambient	air	
- water-cooled isotope production channel	1	Φ 3,6 cm; l. 95 cm	$7 \cdot 10^{17}$	$3,5 \cdot 10^{16}$	$\sim 6,7 \cdot 10^{13}$	100^oC	ca. 1.5 ata	D_2O	Presently used to irradiate UAl fuel plates for production of Te-99 radiopharmaca
- He/D_2O-cooled irradiation channel	1	Φ 7,8 cm; l. 50 cm	$8 \cdot 10^{17}$	$4 \cdot 10^{16}$	$\sim 6,7 \cdot 10^{13}$	100^oC	1,1 bar	He	Installed for irradiation of silicon semiconductor material.
- pneumatic rabbit	1	Φ 3,5 cm; l. 10 cm	$7,2 - 9,5 \cdot 10^{17}$ **)	$5,8 - 6 \cdot 10^{17}$ **)	$\sim 6,7 \cdot 10^{13}$ **)	100^oC	ambient	air	
- thermal column vertical holes	6	20x20x100 cm ***)	$0,3 - 2,2 \cdot 10^{15}$	–	$\sim 10^{8} - 10^{10}$	ca. 100^oC	ambient	air	Revolving sample holding device available for elimination of flux gradient.

*) range of maximal fluxes in these facilities
**) average and maximum flux in this facility
***) last number = axial dimensions within graphite block

Table 3.2 : Irradiation facilities (Cont.)

Type /names	Number	Useful dimensions	Nuclear Characteristics			Environmental Characteristics			Remarks
			Thermal neutron flux in $m^{-2}s^{-1}$	Fast neutron flux in $m^{-2}s^{-1}$ ($E > 0,1$ MeV)	Gamma flux MeV . $cm^{-2}s^{-1}$	Max. allowed temp.	Max. allowed press.	Surr. medium	
<u>Other facilities</u>									
- thermal column horizontal holes	1	60x40x200cm ***)	0,3 - 2,2 . 10^{15}	–	~ 10^{10}	ca. 100°C	ambient	air	The horizontal beam holes and the horizontal holes of the thermal column are occupied by various experiments, preferably in the field of basic neutron physics research. Besides several neutron spectrometers, a cold neutron source is installed and provides a beam of neutrons with energies down to 5 x 10^{-3} eV.
	2	10x10x200cm ***)	,,	,,	,,	,,	,,	,,	
	2	20x20x200cm ***)	,,	,,	,,	,,	,,	,,	
	1	20x20x190cm ***)	,,	,,	,,	,,	,,	,,	
- horizontal beam holes o thimbles	9	Φ in biol. shield: from 22 to 35cm Φ in reflector: from 13 to 29cm	3 - 5 . 10^{17}	1 - 2 . 10^{16}	~ 4 . 10^{13}	ca. 100°C	ambient	air	
o through holes	2	Φ in biol. shield: from 22 to 35cm Φ in reflector and core: 5,5cm	3 - 6 . 10^{17}	1 - 2,5 . 10^{16}	~ 6,7 . 10^{13}	ca. 100oC	ambient	air	
- low-temperature irradiation facility (in core)	1	Φ 1cm; l. 10cm	4 - 5 . 10^{17}	2,3 . 10^{16}	~ 6,7 . 10^{13}	cooling gas temperat. 17 K	4 bar	He	

*) range of maximal fluxes in these facilities
**) average and maximum flux in this facility
***) last number = axial dimensions within graphite block

3.5.2 *Miscellaneous*

Various instrumented capsule experiments for fuel rod irradiation of high-performance fuels, for measurement of fuel creep and swelling under various operating conditions.

Publications:	*Kerntechnik 10 (1968), No. 3; 12 (1970), No. 5/6 and No. 10; 14 (1972), No. 2; 18 (1976), No. 3.*
	Atomwirtschaft 1973, No. 2; 1979, No. 10.
	J. Nucl. Mat. 65 (1977).
	Nucl. Techn. Vol. 34 (1977).

3.5.3 *Pneumatic tube conveyor (rabbit)*

Medium:	*Air under suction*
Operating pressure in the irradiation channel:	*0,7 - 1,0 bar (abs.)*
Air throughput:	*35 m^3/h, velocity 15 m/s*
Irradiation time:	*between ca. 1 and 3.6 . 10^5 s*
ϕ_{th}:	$9 \cdot 10^{17} m^{-2} s^{-1}$
ϕ_f:	$1.5 \cdot 10^{16} m^{-2} s^{-1}$

In addition to the FR 2 building, the isotope laboratory has also a dispatching and receiving unit.

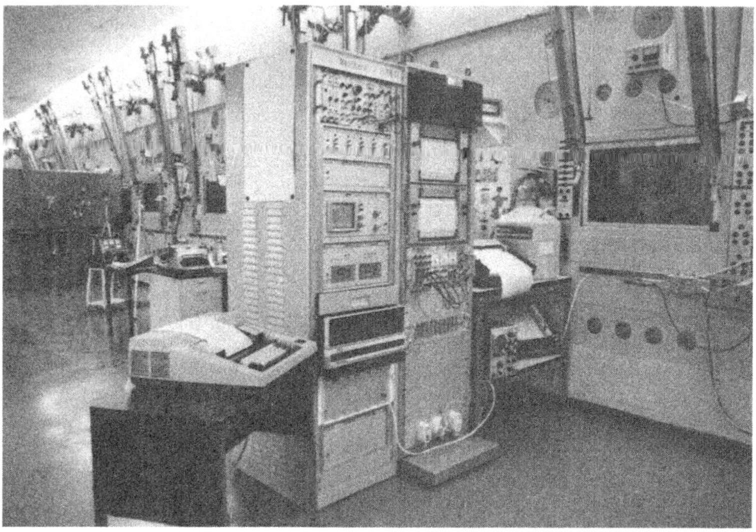

Fig. 3.4 Front view of KfK hot cells line.

3.6 INFORMATION SHEET No. 6. MAIN CHARACTERISTICS OF HOT CELL FACILITIES

Table 3.3: Main characteristics of hot cell facilities

name of laboratory	year of active oper.	scope		gamma shielding (material, thickness, density)	leak tightness	special atmosphere	manipulators number, type (incl. spare manipulators)
		active area (m^2)	number of working positions				
Dismantling cell FR 2 (1)	1961	7	2	concrete	no	no	Mod. 8 GM 300 3 to. crane
Hot cells Building 701	1965	84	11	concrete 90 cm, 4,5 and 3,5 g/cm^3	0,5 vol.⁰/o per h.	N_2	22 Wälischmiller
TPL	1976	60	29	lead, 20 cm	--	N_2	58 Wälischmiller

Remarks: (1) Inspection and dismantling

3.6.1 Owner / operator — Kernforschungszentrum Karlsruhe GmbH, Section Nuclear Plants - Hot Cells

3.6.2 Site — Karlsruhe, KfK

3.6.3 Associated reactor — only the dismantling cell is associated with a reactor (FR 2)

3.7 INFORMATION SHEET No. 7. EQUIPMENT AND TECHNIQUES AVAILABLE FOR POST-IRRADIATION EXAMINATIONS

Dismantling cell, X-ray examination, dimension measurements, gamma spectrometry, gas measuring plant, sample manufacturing, metallography, autoradiography, tensile testing, low-cycle fatigue testing, steady state and transient, creep-rupture testing, internal tube pressure testing, charpy impact testing, hardness testing, X-ray microstructure testing, calorimetry, electron microscopy, scanning microscopy, light microscopy, image analysis, micro sensing, micro-weighing, welding facility.

Relevant publications: "Reports of Hot Cells"

3.8 INFORMATION SHEET No. 8. UTILIZATION AND SPECIALIZATION OF THE HOT CELL FACILITIES

3.8.1 *Original purposes of the facilities*

Post-irradiation investigation on reactor fuel and structural materials.

3.8.2 *Major modifications subsequently made*

Expansion of metallography and structural materials testing.

3.8.3 *Major modifications planned*

Extension of technical investigations.

3.8.4 *Major projects and programmes for which the facilities presently are used*

- *"Fast Breeder" project, 60 %*
- *Reprocessing project, 15 %*
- *Nuclear Safety project, 20 %*

3.8.5 *Availability of the facilities for post-irradiation examination on behalf of other organizations within the European Communities*

At present no free capacity is available for other organizations.

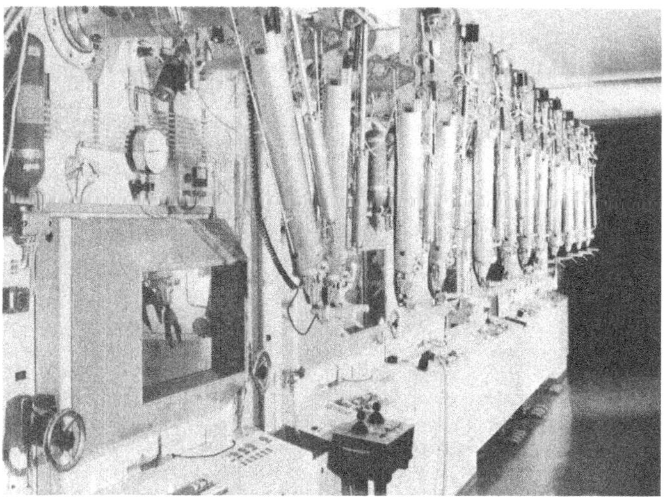

Fig. 3.5 Working positions at the metallography hot cells.

4. THE KNK II REACTOR AT KARLSRUHE, F.R. GERMANY AND AVAILABLE HOT CELL FACILITIES. INFORMATION SHEETS.

Fig. 4.1 ***The KNK II reactor building at the site of KfK Karlsruhe.***

4.1 *INFORMATION SHEET No. 1. MAIN CHARACTERISTICS OF THE REACTOR.*

Name of reactor	*KNK II (Kompakte natriumgekühlte Kernreaktoranlage mit schnellem Kern)*
Organization	*Kernforschungszentrum Karlsruhe GmbH*
Site	*Karlsruhe, Kernforschungszentrum (KfK)*
In operation since	*thermal core 1971-1974; fast core 10.10.1977*
Power	*58* MW_{th} *and 20* MW_{el}
Moderator	*no (only zirconium hydride at the core periphery to protect the vessel wall)*
Coolant	*Na (no pressure)*
Nominal operation (days/year)	*dependent on irradiation and other programmes*
Operating cycle (days) on power and shut down	*dependent on irradiation and other programmes*

Coordinator KNK II test programme (name, address):

Dr. W. Marth Kernforschungszentrum Karlsruhe GmbH
Postfach 3640
D - 7500 Karlsruhe

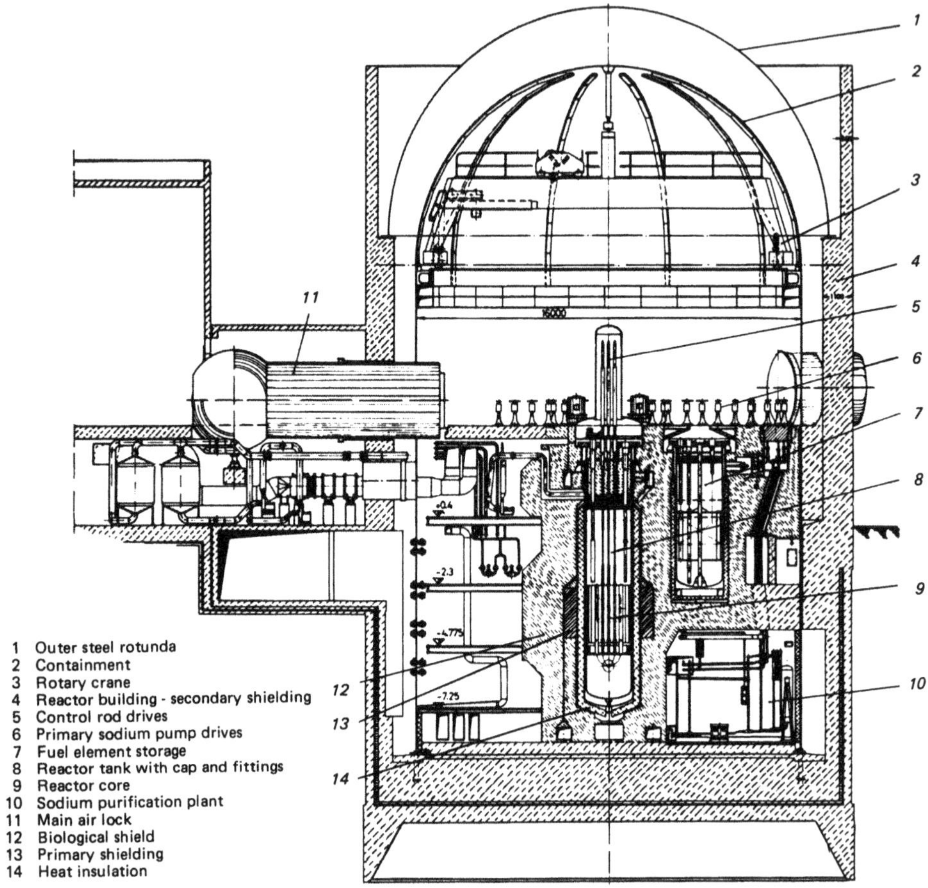

Fig. 4.2 Vertical cross-section of the safety containment.

4.2 INFORMATION SHEET No. 2. UTILIZATION AND SPECIALIZATION OF THE REACTOR

4.2.1 Original purposes of the reactor

- *Irradiation of fuel elements, absorbers and structural materials*
- *Test of reactor and safety instrumentation*
- *Plant tests (e.g. natrium technology)*

4.2.2 Major modifications subsequently made

Rebuilding from thermal to fast reactor.

4.2.3 Major modifications planned

None

4.2.4 Major projects and programmes for which the reactor presently is used

"Fast breeder" Project (PSB)

4.2.5 *Loops and irradiation devices specially developed for the reactor*

Originally the KNK II has been planned as a small-scale fast demonstration reactor. However it is now the intention to use the KNK II in the future more and more as a materials testing reactor to make available its high fast neutron density. Simple non-instrumented capsules have already been developed for the reactor. Other facilities are under construction (see chapter 4.5).

4.2.6 *Additional information*

The reactor vessel, which contains in its lower section the reactor core, is a welded ferritic steel tank, 1,90 m in diameter, 10 m high and 16 mm thick.
The fast KNK core consists of a test zone made up of 7 fuel subassemblies as well as a driver zone consisting of 22 fuel subassemblies. Reactivy control and shutdown are effected by five control rods. In addition, a second shutdown system was provided which is independent of the first shutdown system and consists of three absorber elements.
The core is surrounded by steel-zirconium hydride reflector elements with five breeder elements inserted at the edge of the core (Fig. 4.3 and 4.4).

The fuel elements in the test zone of the first KNK II core include uranium/plutonium mixed oxide pellets as the typical fast breeder fuel. For reasons of reactivity UO_2 has been enriched in U-235. The fuel rod dimensions (outside diameter 6 mm) and the rod spacers in the bundle are in line with the SNR-300 design. In the test zone a maximum power per rod unit length of 600 W/cm (incl. short-term power rises) should be attained, which is the aim for SNR-300.

To ensure the safe removal of afterheat, the nuclear power system has been equipped with two parallel main heat transmission loops of the same capacity. Besides, an emergency cooling system has been provided for KNK II which is completely independent of the function of these loops and makes use of the possibility of direct nitrogen cooling of the reactor tank.
In the main heat transfer system one activity-free secondary sodium circuit each has been provided for safety reasons between the active primary sodium loop and the water vapour loop. In all circuits the sodium is moved by forced convection through mechanical feed pumps. The system generators heated by the secondary sodium are installed outside the safety containment (Fig. 4.2).

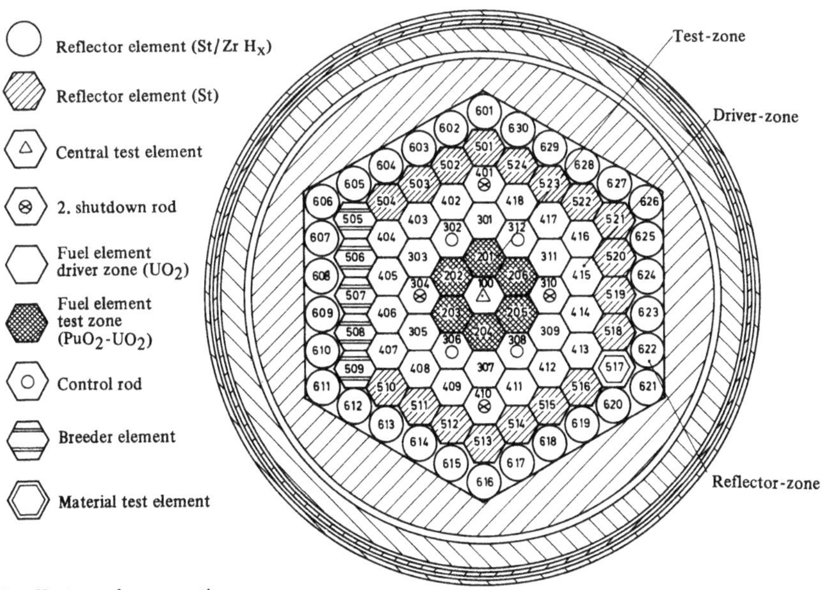

Fig. 4.3 Horizontal cross-section of the KNK II core.

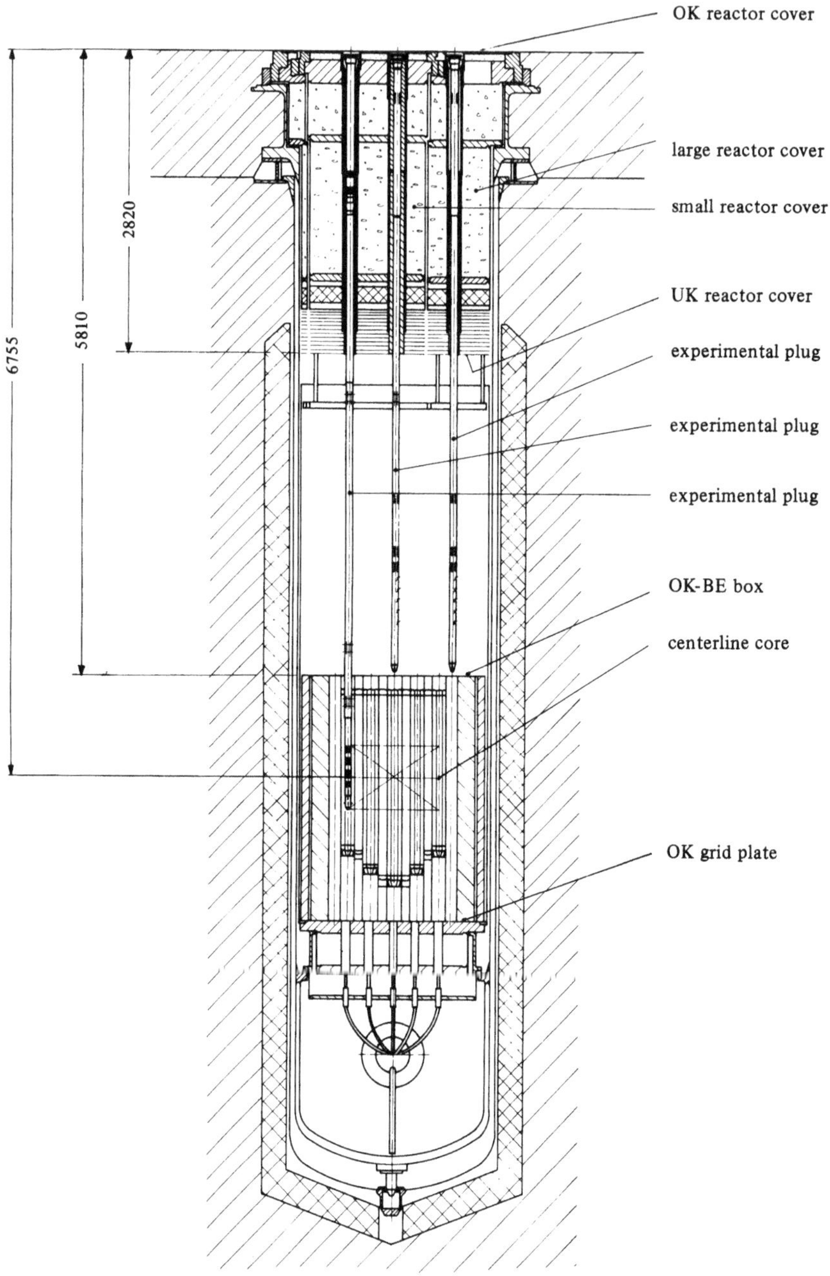

Fig. 4.4 KNK II. Vertical cross-section of the reactor tank.

The main data on the KNK II have been summarized in the following table:

Table 4.1: Main parameters of the KNK II fast reactor

REACTOR CORE

mean core diameter	824 mm
core height	600 mm
number of fuel elements test zone (plutonium/uranium)	7
number of fuel elements driver zone (uranium)	22
number of breeder elements (uranium)	5
number of reflector elements (steel/ZrHx)	30
number of reflector elements (steel)	19
number of rods primary shutdown system	5
number of rods secondary shutdown system	3
fuel inventory, test zone	164 kg Pu_2/UO_2
of which	36 kg Pu_{fiss}
fuel inventory, driver zone	525 kg UO_2
mean power density in test zone/driver zone	183/60 MW/t(heavy metal)
average fast neutron flux in test zone	1.4×10^{19} n/ m^2 s
total neutron flux in test zone	1.67×10^{19} n/ m^2s

PRIMARY CIRCUIT SYSTEM

number of circuits	2
nominal pipe size	200/400 mm
mass flow per circuit	498 t/h
reactor inlet temp.	360 °C
reactor outlet temp.	525 °C
max. pumppressure	5.5 atm.
cover gas	argon

SECONDARY CIRCUIT SYSTEM

number of circuits	2
nominal pipe size	200 mm
flow per circuit	451 t/h
outlet temp. of IHX	504 °C
inlet temp. of IHX	322 °C
max. oper. pressure	13 atm.

INTERMEDIATE HEAT EXCHANGERS

design	cross-counter flow
number	2

STEAM GENERATOR

design	counterflow (double tube coils), water vapour inside, sodium outside
live steam flow	88 t/h

TURBINE

design	single casing condensation turbine
inlet temperature	485 °C
inlet pressure	80 ata

STORAGE POSITIONS

number of positions for storage under sodium	56
number of positions for dry storage	47

4.3 INFORMATION SHEET No. 3 EXPERIMENTAL FACILITIES OF THE REACTOR

4.3.1 Irradiation positions

Table 4.1: Irradiation positions

name	number	active length mm	diameter mm	typical unperturbed fast neutron fluxes in $10^{18} m^{-2} s^{-1}$ (E> 10 keV)
test zone	7	600	102	20
driver zone	22	600	102	10
material test element	1 (2)	600	102	1

4.3.2 Reactivity available for experiment: not applicable

4.3.3 Neutron spectra (data): see chapter 4.4

4.3.4 Specialized equipment available for experimenters:
- experimental rods
- annular base-element in preparation

4.3.5 Availability of facilities for irradiation on behalf of other organizations of the European Communities: Facilities are available

4.3.6 Availability of hot cell facilities: Hot cells are available on site:
- Hot cells of the KfK
- Transplutonium laboratory of the KfK

for details see information sheets on the FR 2 reactor (chapters 3.6-3.8)

4.3.7 Relevant publications:

- W. Marth, H. Mausbeck, H. Andrae, G. Brudermüller
  Errichtung der KNK II-Anlage.
  Atomwirtschaft 3 (1978), p. 118
- W. Kathol
  KNK II - an experimental power station equipped with a fast core.
  Nuclear Engineering International, Jan. 1979, p. 41 - 44
- Kompakte Natriumgekühlte Kernreaktoranlage.
  Special publication by INTERATOM GmbH, Bensberg/Köln

4.4 INFORMATION SHEET No. 4. NEUTRON SPECTRA (DATA)

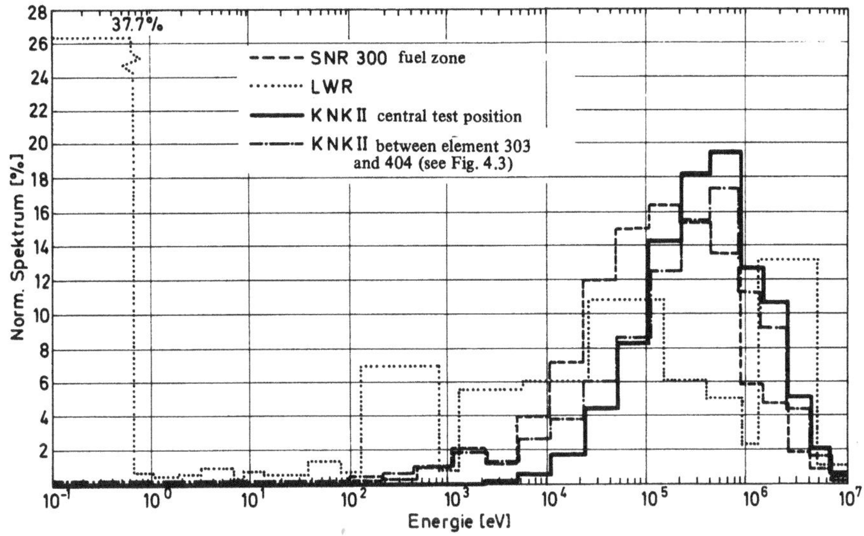

Fig. 4.5 Comparison of neutron spectra for the SNR 300 to those measured for an LWR (1300MW) and for different zones of the KNK II.

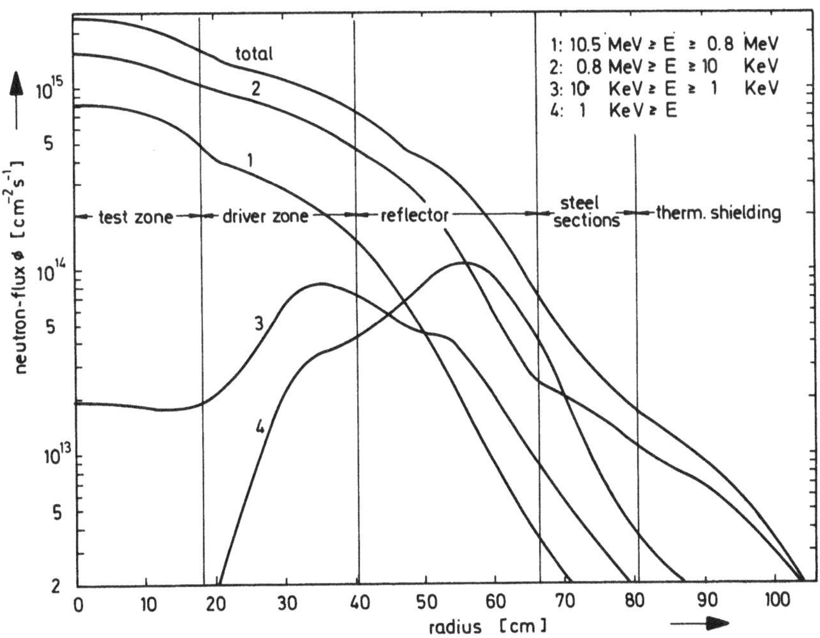

Fig. 4.6 Radial neutron flux distribution in the core central plane (control rod setting 30cm).

4.5 *INFORMATION SHEET No. 5. MAIN CHARACTERISTICS OF SPECIALIZED IRRADIATION DEVICES*

- *non-instrumented capsules for irradiation of cladding and structural material samples in natrium at ambient Na-temperatures (360 - 580 ^{o}C)*
  Useful diameter and length: 6mm and 990mm resp.
- *non-instrumented doublewall capsules for irradiation of pellet loaded fuel pins to study cladding dimensional changes at ambient temperatures.*
  Useful diameter and length: 6mm and 990mm resp.
- *instrumented materials test insert to study in-pile creep behaviour of tubular samples under internal pressure.*
  Internal pressure and the temperature of the samples can be regulated.
  Number of samples: 8
  Length and diameter of the samples: 45mm and up to 9mm resp.
  Surrounding: helium
  Temperature regulation by internal electrical heaters
  Max. internal pressure: 150 bar (constant)
  Insert diameter: 25mm.
- *non-instrumented double-wall Na capsules for irradiation of cladding materials at higher temperatures (550 - 620 ^{o}C), achieved by gas gaps, no temperature control.*
  Temperature measurements by special monitors.
  Useful length and diameter: 600mm and 25mm (outer) resp.

4.6 *INFORMATION SHEET No. 6. MAIN CHARACTERISTICS OF HOT CELL FACILITIES*

See chapter 3.6

4.7 *INFORMATION SHEET No. 7. EQUIPMENT AND TECHNIQUES AVAILABLE FOR POST-IRRADIATION EXAMINATION*

See chapter 3.7

4.8 *INFORMATION SHEET No. 8. UTILIZATION AND SPECIALIZATION OF THE HOT CELL FACILITIES*

See chapter 3.8

5. THE FRJ 1 REACTOR (MERLIN) AT JÜLICH, F.R. GERMANY AND ASSOCIATED HOT CELL FACILITIES. INFORMATION SHEETS.

Fig. 5.1 The FRJ-1 and FRJ-2 reactor buildings at ***KFA Jülich.***

5.1 *INFORMATION SHEET No. 1. MAIN CHARACTERISTICS OF THE REACTOR.*

Name of reactor	*FRJ-1 (MERLIN)*
Organization	*KFA-Jülich GmbH*
Site	*Jülich, Kernforschungsanlage*
In operation since	*1962*
Power MW (th)	*10*
Moderator	H_2O
Coolant	H_2O
Fluid and pressure in experiment channels	*Connected to ventilation system, air with -0.01 bar, resp. coolant water of the primary circuit for vertical channels.*

Nominal operation (days/year) 260

Operating cycle (days)
on power 26
shut down 5

Responsible officers (name, address):

FRJ 1:	*Dr. Friedewold*	*Kernforschungsanlage GmbH*
Hot cells:	*Dr. H. Müller*	*Postfach 1913*
		D - 5170 Jülich

5.2 INFORMATION SHEET No. 2 UTILIZATION AND SPECIALIZATION OF THE REACTOR

5.2.1 Original purposes of the reactor

Research in neutron physics, physics, chemistry, medicine, biology, fuel testing, education and training.

5.2.2 Major modifications subsequently made

- *Substantial modification in 1967 to improve accessibility to reactor core and to run longer operation cycles on power (up to 5 weeks)*
- *Power increase from 5 MW to 10 MW in 1971 to shorten irradiation time.*

5.2.3 Major modifications planned

- *New facility for solid state physics in the A 1 beam tube.*
- *Modification in core area to irradiate large specimen of reactor pressure vessel steel (see also 5.2.4).*

5.2.4 Major projects and programmes for which the reactor presently is used

- *Fuel irradiation for graphite moderated high temperature reactors (especially for testing new reprocessing procedures).*
- *Small sample irradiations for medicine, biology and chemistry.*
- *Irradiation of compact tension blocks (up to $100 \times 240 \times 250\,mm^3$) of material developed for reactor pressure vessels.*

5.2.5 Loops and irradiation devices specially developed for the reactor

- *Several simple irradiation facilities in in-core and edge-positions to allow irradiations with selected neutron fluxes of a broad scale of material (e.g. organic substances) at a temperature level of 50 ^oC.*
- *Fast pneumatic rabbit facility (time for shooting out < 1 s.).*
- *Rotating irradiation facilities for samples of $\leqslant 50\,mm\ \phi$, and $\leqslant 110\,mm\ \phi$ resp.*
- *Three types of irradiation capsules (max. inner dimensions $260 \times 120 \times 500\,mm^3$) for small and large steel samples or other solid materials at a temperature range from 200 to 400 ^oC.*
- *Multi-purpose irradiation rigs for fuel samples cooled by gas or water.*

Main characteristics of specialized irradiation devices are listed in Chapter 5.5.

5.3 INFORMATION SHEET No. 3. EXPERIMENTAL FACILITIES OF THE REACTOR

5.3.1 Irradiation positions

Table 5.1: Irradiation positions

type/name	number **)	active length mm	useful diam. mm	typical unperturbed neutron flux density in $m^{-2}s^{-1}$ thermal	fast (E > 0,1 MeV)	nuclear heating W/g Al
outer core positions *)	24 (7)	600	76,1 x 76,1	0,4 - 0,8 $\times 10^{18}$	1 - 3 $\times 10^{17}$	4,0
in-core positions *)	5 (3)	600	,,	$1 \cdot 10^{18}$	$4 \cdot 10^{17}$	6,4
horizontal beam tubes						
D3, E5, F6	3	1000	153 Φ	$7 \cdot 10^{16}$	$5 \cdot 10^{15}$	--
A1	1	1000	300 Φ	,,	,,	
thermal columns						
THI, channels 1 through 4, and THII, channels 1 through 3	7	1500	101 x 101	$2 \cdot 10^{14}$	$2 \cdot 10^{12}$	--
vertical channels						
V1, V2	2	200	150 Φ	$2 \cdot 10^{16}$	--	--
Rabbit irradiation facilities						
BE 25 ***)	1	30	10 Φ	$4 \cdot 10^{17}$	$6 \cdot 10^{16}$	
BE 22	5	110	10-50 Φ	$1 \cdot 10^{18}$	$4 \cdot 10^{17}$	4,0
BE 1	1	100	23 Φ	$5 \cdot 10^{17}$	$1 \cdot 10^{17}$	
BE 26D	1	300	50 Φ	$4 \cdot 10^{17}$	$1 \cdot 10^{17}$	
BE 43	1	300	110 Φ	$1 \cdot 10^{17}$	$5 \cdot 10^{16}$	

*) vertical tubes on lattice
**) Numbers in brackets available after core modifications (see 5.2.3)
***) see Table 5.2

5.3.2 Reactivity available for experiments: $\Delta k/k = 2{,}5$ to 9,5 °/o, depending on burn-up of fuel

5.3.3 Neutron spectra (data): see chapter 5.4

5.3.4 Specialized equipment available for experimenters: storage block and radiation protection flasks for handling long and short experiment; experiments facility; neutrography (small unit) $10^5 cm^{-2}s^{-1}$; small hot cells for remote handling of isotopes up to 2 Curie (1 MeV).

5.3.5 Availability of facilities for irradiation on behalf of other organizations of the European Communities: Facilities are available.

5.3.6 Availability of hot cell facilities: Hot cell facilities are available on the reactor site (see chapter 5.6).

5.3.7 *Relevant publications:*

- *Sicherheitsbericht des Reaktors FRJ-1 (MERLIN) vom 26. April 1961.*
- *Sicherheitsbericht vom. 5. August 1963 (addendum to the above-mentioned report)*
- *Sicherheitsbericht für den Betrieb des Forschungsreaktors FRJ-1 (MERLIN) mit einer thermische Leistung von 10 MW.*

Fig. 5.2 FRJ-1. View into the reactortank.

5.4 INFORMATION SHEET No. 4. NEUTRON SPECTRA (DATA)

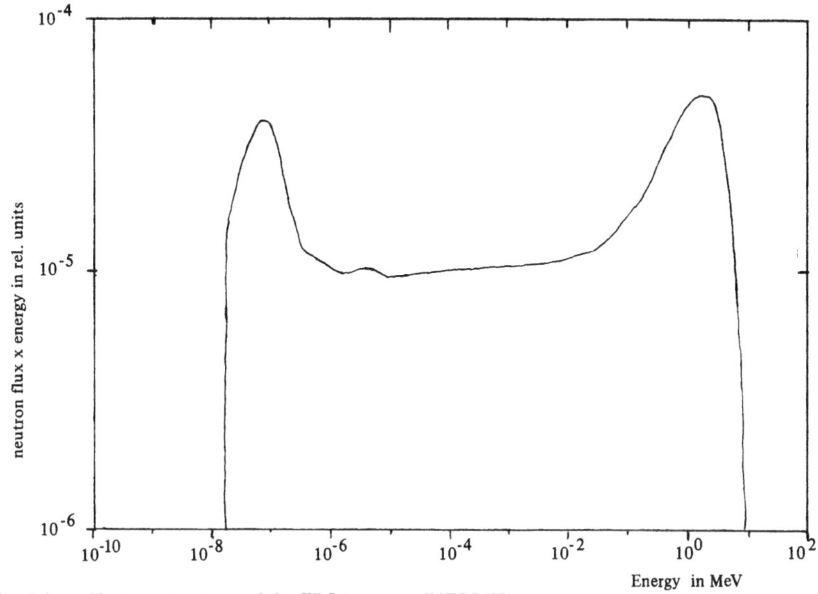

Fig. 5.3 Neutron spectrum of the FRJ-1 reactor (MERLIN).

5.5 INFORMATION SHEET No. 5. MAIN CHARACTERISTICS OF SPECIALIZED IRRADIATION DEVICES.

5.5.1 Main characteristics are summarized in Table 5.2

5.5.2 Relevant publications:

- *W. Marth*
  Bestrahlungstechnik an Forschungsreaktoren.
  Thiemig Verlag, München

- *M. Pollermann*
  ZIR-Ergebnisse und Möglichkeiten.
  KFA-Bericht, JÜL-796-RX

- *M. Pollermann*
  The facilities for reactor irradiation.
  Special issue KFA Jülich, 1964.

- *D. Pachur*
  Rigbestrahlungseinrichtungen an den Reaktoren.
  Atompraxis 10 (1964)

- *D. Pachur*
  Rigbestrahlung von Stahlproben.
  Kerntechnik 6 (1964)

- *H. Stechemesser, K. Sachse.*
  Hydraulische Rohrpost.
  Kerntechnik 12 (1970)

- *G. Borchardt, et al.*
  Ergebnisse des Flussdichtemessungen in Experimentepositionen des FRJ-1 und FRJ-2 nach Leistungserhöhung auf 10 MW bzw. 25 MW.
  KFA-Bericht, JÜL-1344, 1976

- *W. Scharenberg, H.J. Bormann*
  Neutronenradiographische Untersuchungen am Reaktor FRJ-1.
  Kerntechnik 18 (1976).

- *A. Butzek, et al.*
  Entwicklung, Bau und Betrieb eines Rig-Brennelementes für Vertikalrigs im FRJ-1.
  KFA-Bericht, JÜL-1480, 1978.

Table 5.2: Main characteristics of specialized irradiation devices

device type	position characteristics	number	useful dimensions diameter/length (mm)	nuclear characteristics: thermal neutron flux density ($m^{-2}s^{-1}$)	nuclear characteristics: fast neutron flux density ($m^{-2}s^{-1}$) ($E > 0,1\,MeV$)	gamma flux r/h	environmental characteristics: capsule temperature ^oC	environmental characteristics: surrounding medium
BE 3	in-core position	1	50 / 600	$1 . 10^{18}$	$4 . 10^{17}$	10^9	50	H_2O
		5	50 / 600	$2 . 10^{17}$	$3 . 10^{17}$	$2 . 10^8$	50	H_2O
BE 22	tube insertion procedure (max. flux level position)	1	50 / 110	$1 . 10^{18}$	$4 . 10^{17}$	10^9	50	H_2O
	tube insertion procedure (upper level)	1 *)	50 / 110	$8 . 10^{17}$	$3 . 10^{17}$	$2 . 10^8$	50	H_2O
	tube insertion procedure (medium level)	1 *)	50 / 110	$7 . 10^{17}$	$2 . 10^{17}$	$2 . 10^8$	50	H_2O
	tube insertion procedure (lower level)	1 *)	50 / 110	$6 . 10^{17}$	$1 . 10^{17}$	$2 . 10^8$	50	H_2O
BE 1	pneumatic rabbit	1	23 / 100	$5 . 10^{17}$	$1 . 10^{17}$	$2 . 10^{18}$	110	CO_2
BE 25	fast rabbit	1	10 / 30	$4 . 10^{17}$	$6 . 10^{16}$	$2 . 10^8$	110	CO_2
BE 26 D	capsule rotating facility	1	50x50 / 300	4×10^{17}	$1 . 10^{17}$	10^8	50	H_2O
BE 15	cage facility (thermal column)	1	65x55 / 15 **)	$2 . 10^{13}$	10^{11}	$2 . 10^2$	40	air
BE 14	low temp. irradiation (thermal column)	1	200 x 60 / 60	$1 . 10^{14}$	10^{12}	$6 . 10^3$	down to 0	air / CO_2
BE 43	capsule rotating facility	1	111 / 300	$1 . 10^{17}$	$5 . 10^{16}$	$1 . 10^8$	50	H_2O
BE 9.8	CT-capsule	1	120 x 260 / 500	$5 . 10^{16}$	$2 . 10^{17}$	$7{,}5 . 10^8$	200 to 400	He

*) 3 samples holders per unit
**) dimensions for small flux gradients

5.6 INFORMATION SHEET No. 6. MAIN CHARACTERISTICS OF HOT CELL FACILITIES

Table 5.3: Main characteristics of hot cell facilities

name of labor.	year of active oper.	scope		gamma shielding (material, thickness, density)	leak tightn.	special atmosph.	manipulators number, type (incl. spare manipulators)
		active area (m^2)	number of working pos.				
HZ 1 (refs. 1, 2)	1969	70 (cell) 95	11	barytic concrete 110 cm, 85 cm 3.5 g/cm^3	a, γ	N_2	8 Draht & Schrader ELMA 50-1000-2 24 HWM 100 1 HWM 200
HZ 2 (refs. 1, 2)	1970	52	15	barytic concrete 85 cm, 3,5 g/cm^3	a, γ	N_2	2 Draht & Schrader ELMA 50-1000-2 30 HWM 100
HZ 3	1980	6	2	lead, 20 cm	a, γ	--	4 HWM 100
BZ 1 *) (ref. 3)	1965	10	7	lead, 25 cm lead, 15 cm lead, 10 cm	β, γ	--	3 CRL 7 17 (rod grabs)
BZ II (ref. 3)	1966	15,5	7	cast iron, 46 cm 7,0 g/cm^3	a, γ boxes	--	12 HWM 100 1 CRL
BZ III (ref. 3)	1967	18,5	7	concrete, 140 cm 2,3 and 2,8 g/cm^3 iron, 40 cm, 7,5 g/cm^3 lead, 10 cm	a, γ boxes	--	9 HWM 100 1 HWM 200 2 CRL 7
"Chemie-zellen"	1973	51 4 35	15 hot cell - boxes 2 transfer stations 1 techno-logical box with variable shielding	barytic concrete 85 cm, 3,5 g/cm^3 lead, 25 cm and 20 cm lead, variable	a	none	40 Master-slave manipulators, Wälischmiller A-100 type

*) BZ 1 cells are being reconstructed.

References: H. O. Witte, H. Müller
Proc. 17th Conf. Remote Syst. Technology (1969), p. 87 - 98.

H. Müller, K. Stradel, H. Witte
Atomwirtschaft XIII, 8-9 (1968), p. 444 - 448.

R. Buch, M. Herren, H. Müller
Safety Report, KFA-IRW-IB-29/79

5.6.1	*Owner / operator*	*Kernforschungsanlage Jülich GmbH*
5.6.2	*Site*	*Jülich, KFA*
5.6.3	*Associated reactor*	- *HZ I, HZ II and HZ III are associated with FRJ-1, FRJ-2 and AVR (connection only by means of motor vehicles)* - *BZ I is not directly associated to a reactor. Transfer of samples for PIE from other cells of KFA.* - *BZ II and BZ III are associated with FRJ-1, FRJ-2 and AVR (Jülich); BR2 (Mol); HFR (Petten); Halden reactor; Cadarache; Studsvik (connection only by motor vehicles with shielding containers).* - *"Chemiezellen" are associated with FRJ-1 and AVR.*

5.7 INFORMATION SHEET No. 7. EQUIPMENT AND TECHNIQUES AVAILABLE FOR POST-IRRADIATION EXAMINATIONS

BZ I

The BZ I is mainly equipped for the examination of irradiated structural reactor materials.
Tensile test, compression test up to 100 kW.
Temperature gradient test for coated particles up to 1800 °C in argon atmosphere.
X-raying of small samples.
Determination of the crystalline proportion in notched-bar impact test samples, photography.
Grinding notched-bar impact tests, hardness test. Length marking, length measurement, weighing.

BZ II

Dismantling of reactor modules and irradiation capsules; longitudinal and transversal sectioning, removal of test samples, repacking, periscope photography, length measurement, machining of tensile test specimens.
Gamma scanning up to a length of 1 m, gamma spectrometry, measurements of spherical-element burn-up (Cs determination)
Diameter and generatrix measurement (1 m long, 80 mm in diameter), optical inspection, puncturing gas pressure measurement and removal of fission gas, void determination.
Preparing of test samples without cross contamination for fission product behaviour studies.

BZ III

Test sample removal, disassembling of irradiation capsules, optical inspection and periscope photography, dimension measurements, weighing, oxidation test.
Annealing test of sperical fuel elements and coated particles up to 250 °C mainly with regard to reactor failing conditions.
Gamma spectrometry and gamma scanning up to a specimen length of 0,5 m.
Studies of long term storage behaviour for HTR-fuel with regard to fission gas release.
Metallography and micro hardness testing, beta-autoradiography, detection of defect-rate for coated particles.

HZ 1 (incl. HZ 3, outside cells)

Inward transfer of reactor modules (horizontal and vertical up to 400mm diameter, dismantling, leak testing.
Gamma scanning of fuel rods up to 2 m length.
Turning, milling, abrasive cutting, capsule opening, repacking, gamma spectrometry of spherical elements.
Backing-out testing of coated particles, drop testing of spherical elements, weighing.
Test sample storage.
Metallographs and micrographs of ceramics, removal of micro-specimens, miscroscopy.
Up to circa 1980, 60.000 AVR spherical fuel elements will be stored in the cooling pond of the HZ 1.

HZ 2 *The laboratory is now free of contamination.*
A reprocessing plant on a semi-works scale (graphite combustion, dissolution of the fuel) for spherical THTR fuel elements is under installation.
The laboratory will thus be occupied until about 1988.

HZ 3 *Electron scan microscopy of small, and also radioactive, bodies, such as coated particles, powders, etc. (outside the cells).*
Electrolytic desintegration of spherical fuel elements and compacts.
Gamma spectrometry of resulting samples.

"Chemiezellen":

The gastight boxes in non-gastight cells of barytic concrete are interchangeable.
Each experiment is assembled inside a box and cold-tested in the mock-up room, and the fully-equipped box is then moved into the charging zone of the concrete cells. The fully-equipped connection module for the electrical supply and control of the experiment is inserted into the appropriate console situated in the service area, so that no further electrical connection need to be made at the site of the experiment.
In the charging zone the box is taken over by the manipulator, the shielding door of the charging zone is closed, and the box is placed by remote control on two forks which are attached to the plug-door. When the plug-door is closed, the box is moved to a position above the specially designed stand, and on being lowered into this stand is automatically coupled to all the supply connections.
When a box is being removed, the plug-door performs the reverse function of uncoupling the box from the supply connections and returning it to the charging zone.
All the boxes in both cell banks are gastightly connected to an underlying conveyor tunnel in which a dolly runds and from which they are sealed off by a cover inside the boxes.
All transport of materials and radioactive substances from the transfer station to the individual boxes is effected via this tunnel. Besides commonly used transport systems, the PADIRAC and the CENDRILLON systems will be used for the transport of radioactive solids and radioactive liquids, respectively.

Relevant publications

- *W. Schenk*
  Nachbestrahlungsausheizverfahren für Kugelbrennelemente and andere Brennstoffproben.
  KFA-report (JÜL-1454), 1977.

- *W. Schenk*
  Untersuchungen zum Verhalten von beschichteten Brennstoffteilchen und Kugelbrennelementen bei Störfalltemperaturen.
  KFA-report (JÜL-1490), 1978.

- *G. Kaiser, P. Sckuhr, H. Witte*
  Planning of a semi-works scale experimental plant for the reprocessing of HTGR fuel elements.
  Paper presented at the meeting on reactors of the German Atom Forum, Bonn, 30 March - 2 April, 1971.

- *H. Müller, K. Stradal, H. Witte*
  Work programme and equipment of hot cells 1 and 2 in Jülich.
  Atomwirtschaft XIII, no. 8-9 (1968), p. 444 - 448.

- *K.A. Stradal*
  Disassembling machine for in-core thermionic diodes.
  Post-Irradiation Examination Techniques, Conference, Reading (U.K.), March 22 - 23, 1972.
  London, The Inst. of Civil Engineers, Vol. C, p. 43 - 44.

- *J.M. Davis, B. Javary, A. Tarbe de St.Hardouin and L. Schlösser*
  Proc. 13th Conf. Remote Syst. Techn. (1965), p. 69.

- *H. Müller, F. Stockschlaeder, H. Witte*
  Radiation Safety in Hot Facilities; pp. 193 - 200; IAEA, Vienna, 1970.

- *H. Müller, F. Stockschlaeder, H. Witte*
  Radiation Safety in Hot Facilities; pp. 461 - 472; IAEA, Vienna, 1970.
- *H.J. Riedel, W. Utecht*
  Report JÜL-777-CT, July 1971
- *St. Halaszovich*
  Verfestigung hochradioaktiver Spaltproduktlösungen aus der Wiederaufarbeitung abgebrannter Brennelemente für eine Endlagerung.
  JÜL-1636, Jan. 1980.
- *Fachseminar zur WA und Entsorgung im Thorium-Brennstoffkreislauf; pp. 104 - 116, pp. 141 - 175.*
  JÜL-Conf. 30, Febr. 1979.

5.8 INFORMATION SHEET No. 8. UTILIZATION AND SPECIALIZATION OF THE HOT CELL FACILITIES

5.8.1 Original purposes of the facilities

BZ I	Steel studies
BZ II	Test fuel-rod studies
BZ III	Studies of THTR fuel elements (spheres)
HZ 1	General materials testing of large components
HZ 2	Physical and chemical studies of all kinds
HZ 3	Area prepared for setting up of lead cells
"Chemiezellen"	Chemical work on hot radioactive materials

5.8.2 Major modifications subsequently made

BZ I	--
BZ II	--
BZ III	--
HZ 1	Conversion to PADIRAC airlock system
HZ 2	Conversion of the area to the maintenance of α-tight boxes Changeover of the airlock system
HZ 3	2 Lead cells are installed
"Chemiezellen"	

These conversions were carried out, without any particular programme in mind, simply to keep pace with the latest developments in technology.

5.8.3 Major modifications planned

None

5.8.4 Major projects and programmes for which the facilities presently are used

- *HBK-Project; safety of nuclear facilities*
- *"Entsorgung" AVR fuel elements*
- *The "Chemiezellen" will be used in the reprocessing programme which is the main project of the Institute for Chemical Technology. The work includes experiments for waste treatment, refabrication of fuel particles, testing of experimental components under hot conditions and all the hot analytical work in connection with JUPITER (JUelicher PIlot-Anlage für Thorium-Element Reprocessing).*

5.8.5 Availability of facilities for post-irradiation examinations on behalf of other organizations within the European Communities

All the hot laboratories mentioned above are in a position to accept projects that fall within the scope and capacity of their equipment, provided their work schedule permits it; (HZ 2 will be occupied until 1988, the "Chemiezellen" have no free capacity at present).

6. THE FRJ 2 REACTOR (DIDO) AT JÜLICH, F.R. GERMANY AND ITS ASSOCIATED HOT CELL FACILITIES. INFORMATION SHEETS.

6.1 *INFORMATION SHEET No. 1. MAIN CHARACTERISTICS OF THE REACTOR.*

Name of reactor	*FRJ-2 (DIDO)*
Organization	*KFA-Jülich GmbH*
Site	*Jülich, Kernforschungsanlage (See Fig. 5.1)*
In operation since	*1962*
Power MW (th)	*23*
Moderator	D_2O
Coolant	D_2O
Fluid and pressure in experiment channels	*Connected to ventilation system air with - 0,01 bar.*
Nominal operation (days/year)	*260*
Operating cycle (days)	
on power	*23*
shut down	*5*

Responsible officers (name, address):

FRJ 2:	*Dr. Friedewold*	*Kernforschungsanlage GmbH* *Postfach 1913* *D - 5170 Jülich*

6.2 *INFORMATION SHEET No. 2 UTILIZATION AND SPECIALIZATION OF THE REACTOR*

6.2.1 *Original purposes of the reactor*

Materials testing at 10 MW thermal power

6.2.2 *Major modifications subsequently made*

- *Power increase to 15 MW (1967)*
- *Power increase to 23 MW (1972)*
- *Cold water cooling circuit for experiments*
- *Better working conditions*

} *generally for all experiments*

6.2.3 *Major modifications planned*

New experimental facilities for solid state physics.

6.2.4 *Major projects and programmes for which the reactor presently is used*

- *Beam experiments for solid state physics*
- *Beam experiments for neutron physics*
- *Irradiation experiments for HTR- and BWR-fuels*
- *Irradiation of steel samples*

6.2.5 *Loops and irradiation devices specially developed for the reactor*

- *Liquid hydrogen loops for low energy neutrons*
- *Boiling water loop facility for testing canned fuel rods under irradiation conditions (pressure up to 130 bar) (4 loops at the same time).*
- *Irradiation rigs for coated particles in connection with sampling facility to measure release of fission products.*
- *4 flight spectrometers, one of them with backscattering devices*

Main characteristics of specialized irradiation devices are listed in chapter 6.5.

6.3 INFORMATION SHEET No. 3. EXPERIMENTAL FACILITIES OF THE REACTOR

6.3.1 Irradiation positions

Table 6.1: Irradiation positions

type/name	number	active length mm	useful diam. mm	typical unperturbed neutron flux density in $m^{-2}s^{-1}$ thermal	fast (E > 0,1 MeV)	nuclear heating W/g Al
fuel element (vertical)	25	610	51	$2 \cdot 10^{18}$	$1 \cdot 10^{18}$	2,5
vertical in D_2O 2V, 4V, 6V	7/4/4	610	51/102/153	2 - 0,5 $\times 10^{18}$	$1 \cdot 10^{17}$	0,3
vertical in graphite 4 VGR, 6 VGR, 10 VGR	2/6/2	610	102/153/254	$1 \cdot 10^{17}$	$1 \cdot 10^{14}$	0,005
horizontal in D_2O 4H, 6H, 10H	6/1/1	1500	108/152/253	$1 \cdot 10^{18}$	$1 \cdot 10^{15}$	0,7-1,5
horizontal in graphite 6 HGR	10	1100	152	$2 \cdot 10^{17}$	$2 \cdot 10^{14}$	0,025
horizontal, thermal column 4 HTC	9	1100	101	$7 \cdot 10^{14}$	10^{13}	1,15
below the core in D_2O 2 TAN	1	2000	21,5/85	$2 \cdot 10^{18}$	$1 \cdot 10^{17}$	1,15

Fig. 6.1 FRJ-2 (DIDO). View into the reactor hall.

6.3.2	*Reactivity available for experiments:*	*8 °/o Δk/k (about 10 irradiation facilities in fuel elements and V-channels.)*
6.3.3	*Neutron spectra (data):*	*see chapter 6.4*
6.3.4	*Specialized equipment available for experimenters:*	*Cold neutron source (liquid hydrogen)*
6.3.5	*Availability of facilities for irradiation on behalf of other organizations of the European Communities:*	*Facilities are available.*
6.3.6	*Availability of hot cell facilities:*	*Hot cell facilities are available on the reactor site (see chapters 5.6, 5.7 and 5.8).*

6.3.7 *Relevant publications:*

- *Safety report on research reactor FRJ-2 (DIDO), Febr. 1962.*
- *Safety report on the operation of FRJ-2 at 15 MW, May 1967.*
- *Safety report on the operation of FRJ-2 up to a maximum of 25 MW, Oct. 1970.*

6.4 *INFORMATION SHEET No. 4. NEUTRON SPECTRA (DATA)*

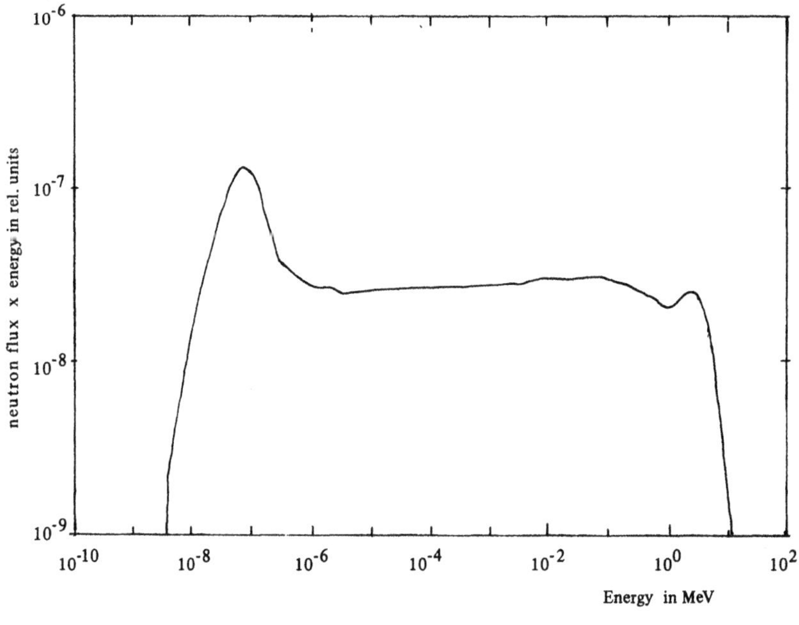

Fig. 6.2 Neutron spcetrum of the FRJ-2 reactor (DIDO).

6.5 INFORMATION SHEET No. 5. MAIN CHARACTERISTICS OF SPECIALIZED IRRADIATION DEVICES

Table 6.2: Main characteristics of specialized irradiation devices

device	number	useful dimensions diameter/length (mm)	nuclear characteristics		gamma flux r/h	environmental characteristics	
			thermal neutron flux density $(m^{-2}s^{-1})$	fast neutron flux density $(m^{-2}s^{-1})$ $(E > 0.1\ MeV)$		capsule temperature ^oC	surrounding medium
BE 9 / LV 48 Multi Purpose Irradiation Device	6	10 x 10 x 55 (24 in each)	$1{,}5 . 10^{18}$	$4 . 10^{17}$	$2 . 10^9$	250 - 700	He/Ne
BE 21 Multi Purpose Irradiation Device	2	45 x 20 (2 in each) 12 x 200 (8 in each)	$1{,}5 . 10^{18}$	$4 . 10^{17}$	$2 . 10^9$	70	D_2O
BE 20 Hydraulic Rabbit	1	23 x 80	$1{,}4 . 10^{18}$	$1 . 10^{17}$	$2 . 10^9$	4	H_2O
BE 12 Standard Rabbit	1	23 x 80 (2 in each)	$3{,}5 . 10^{17}$	$1{,}5 . 10^{15}$	$8 . 10^7$	100	CO_2
BE 11 Fast Rabbit	1	4 x 25	$6 . 10^{17}$	$5 . 10^{14}$	$2 . 10^8$	200	CO_2
Isotope Unit	1	23 x 70 (70 in each)	$10^{14} - 10^{17}$	$10^{11} - 10^{13}$	10^7	40 - 150	CO_2
Fuel Irradiation Loops in Core Automatic Analysing System for Fission Care			$1{,}4 . 10^{18}$	$4 . 10^{17}$	$2 . 10^9$	400 - 600	He/Ne

6.5.1 *Relevant publications*

- *F. Schmidt*
  Die Entwicklung von Brennstoff- und Strukturmaterial-Brennelementen unter Anwendung des Baukastenprinzips.
  KFA-report (JUL-828-RX), 1972.

- *A. Butzek, H. Küpper*
  Axiale und radiale Flussverteilung sowie nukleare Aufheizraten in verschiedenen Brennelementpositionen nach Leistungserhöhung des FRJ-2 auf 25 MW, gemessen in BE 9.3-Einschüben.
  KFA-report (JUL-1048-RX), 1973

- *L. Weise*
  Prognose der schnellen Neutronenflussdichte in Brennelement-Kanälen des Forschungsreaktors FRJ-2.
  KFA-report (JUL-1084-BB), 1974.

See also chapter 5.5.1

6.6 *INFORMATION SHEET No. 6. MAIN CHARACTERISTICS OF HOT CELL FACILITIES*

See chapter 5.6

6.7 *INFORMATION SHEET No. 7. EQUIPMENT AND TECHNIQUES AVAILABLE FOR POST-IRRADIATION EXAMINATIONS*

See chapter 5.7

6.8 *INFORMATION SHEET No. 8. UTILIZATION AND SPECIALIZATION OF THE HOT CELL FACILITIES*

See chapter 5.8

7. THE FRG 2 REACTOR AT GEESTHACHT, F.R. GERMANY AND ITS ASSOCIATED HOT CELL FACILITIES. INFORMATION SHEETS.

Fig. 7.1 Air view of the Geesthacht research centre.

7.1 INFORMATION SHEET No. 1. MAIN CHARACTERISTICS OF THE REACTOR.

Name of reactor	*Forschungs-Reaktor-Geesthacht-2 (FRG-2)*
Organization	*GKSS-Forschungszentrum Geesthacht GmbH*
Site	*Geesthacht-Tesperhude*
In operation since	*1963*
Power MW (th)	*15*
Moderator	H_2O
Coolant	H_2O
Fluid and pressure in experiment channels	--
Nominal operation (days/year)	*280*
Operating cycle (days)	
on power	*35*
shut down	*5*

Responsible officer (name, address):

Dr. G. Röbert *GKSS-Forschungszentrum Geesthacht GmbH*
Zentralabteilung Forschungsreaktoren
D - 2054 Geesthacht

Fig. 7.2 View of the reactor hall; front FRG-2; back: FRG-1.

7.2 INFORMATION SHEET No. 2 UTILIZATION AND SPECIALIZATION OF THE REACTOR

7.2.1 Original purposes of the reactor

Nuclear research and materials testing

7.2.2 Major modifications subsequently made

- *Back fitting in 1978 / 79*

7.2.3 Major modifications planned

- *Reduction of enrichment in U235 in MTR-fuel elements.*

7.2.4 Major projects and programmes for which the reactor presently is used

- *Research programme: "Integrity of components" ; embrittlement of RPV steels.*
- *Creep of Zircaloy fuel cans.*

7.2.5 Loops and irradiation devices specially developed for the reactor

- *Magazine rigs for structural materials*
- *Creep capsule for structural materials*
- *Water corrosion capsule*
- *Container capsules for large pressure vessel steel specimens*
- *Irradiation rig operating at PWR conditions for investigating defected fuel rods*
- *Irradiation device for the determination of in-pile creep behaviour of zircaloy cladding tubes under internal and external overpressure*

for main characteristics of specialized irradiation devices see chapter 7.5

Fig. 7.3 FRG-2; view of the reactor pool.

7.3 INFORMATION SHEET No. 3. EXPERIMENTAL FACILITIES OF THE REACTOR

7.3.1 Irradiation positions

Table 7.1: Irradiation positions

type/name	number	active length mm	diameter mm	typical unperturbed neutron flux density in $m^{-2}s^{-1}$ thermal	fast ($E > 0.1$ MeV)	nuclear heating W/g
In-core positions	3	600	75	1.2 - 1.5 x 10^{18}	0.4 - 1.2 x 10^{18}	1.8 . 10^{9} *)
Peripheral positions	20	600	75	0.3 - 1.0 x 10^{18}	1.5 - 4.5 x 10^{17}	8 . 10^{8} *)

*) related to Al in rd/h

7.3.2 *Reactivity available for experiments* — *No limitation if the experiments are part of the reactor core*

7.3.3 *Neutron spectra (data):* — *Typical neutron spectra of a light water MTR-type research reactor, see chapter 7.4.*

7.3.4 *Specialized equipment available for experimenters:* — *Neutron radiography*

7.3.5 *Availability of facilities for irradiation on behalf of other organizations of the European Communities:* — *Facilities are available*

7.3.6 *Availability of hot cell facilities:* — *Hot cell facilities are available on the reactor site, see chapter 7.6 through 7.8.*

7.3.7 *Relevant publications.*

- J. Ahlf and G.A. Röbert
  The research reactors and irradiation facilities at the Research Centre Geesthacht of the G.K.S.S.
  (72,I,12) EAES Symposium at Petten, 16 - 17 May, 1972.
- G.A. Melkonian (Ed.)
  Instrumentierte Bestrahlungseinrichtungen für Reaktorwerkstoffe am Forschungsreaktor Geesthacht (FRG-2).
  KEST-Bericht Nr. 26, München, 1975.

7.4 INFORMATION SHEET No. 4. NEUTRON SPECTRA (DATA)

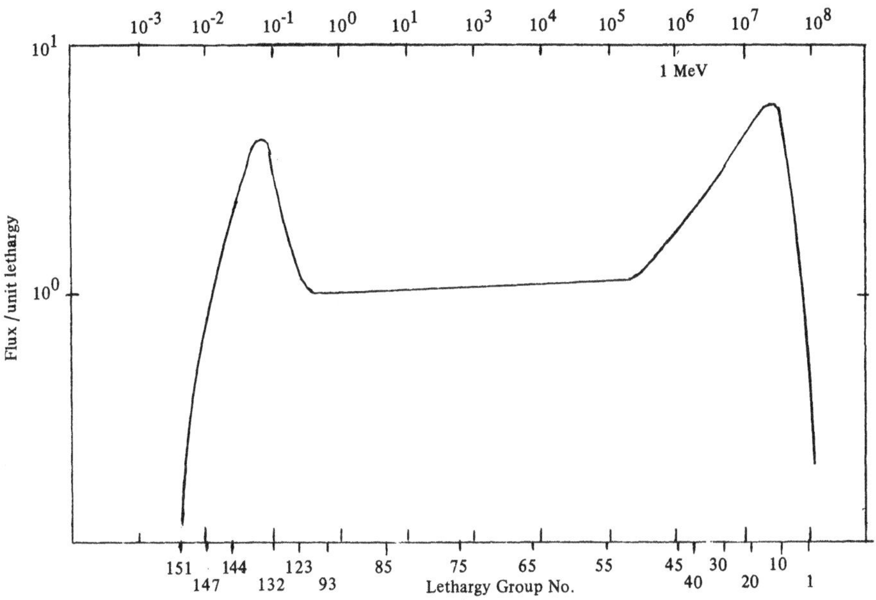

Fig. 7.4 FRG-2 (and FRG-1) reactor, typical neutron spectra

7.5 INFORMATION SHEET No. 5. MAIN CHARACTERISTICS OF SPECIALIZED IRRADIATION DEVICES.

7.5.1 Irradiation Devices

- *Magazine Rig A5:* *for non fissile materials, 200 - 1000 °C, electrical heating 3 kW, up to 15 bar He, Ne, N_2 mixtures, 100 W/cm².*
- *Magazine Rig F1:* *inert gas mixture 5 bars for temperature control, specimen diameter 20mm; length 240 - 480mm, 500 W/cm.*
- *Creep Rig K1:* *5 tubular specimens, 350 - 450 °C, in He-Ne gas mixture, maximum force 4000 N.*
- *Creep Rig K3:* *6 tubular specimens, 500mm long, 280 - 400 °C, biaxial tensile or compressive stresses from 70 to 150 N/mm², temperature control by forced convection of He.*
- *Water Corrosion Rig W3:* *for Zircaloy cans at full PWR conditions, at 350 °C and 150 bars pressure.*
- *Device H T 1:* *for HTGR fuel irradiation; 600 - 1400 °C, on line fission gas analysis (out of operation).*
- *Container Capsule CK:* *specimen volume up to 120 x 260 x 500mm³, 250 - 350 °C, He-N_2-mixture; temperature control by multi-zone electrical heating.*

LWR Fuel Rod Rig Z1: *PWR conditions (160 bar, 320 °C, water chemistry control), 1 fuel rod, active length 300mm, max. linear power 800 W/cm² (under development).*

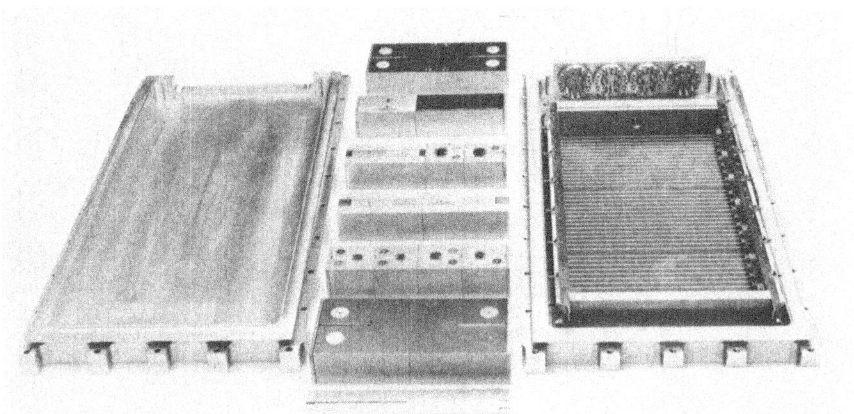

Fig. 7.4 Container capsule for large pressure vessel steel specimens.

7.5.2 *Relevant Publications*

- *J. Ahlf*
  Irradiation rig operating at PWR conditions for investigating defected fuel fods.
  GKSS 79/E/51 (1979)
- *J. Ahlf, A. Reymann, O. Eichborn, M. Gärtner*
  Design of an irradiation device for the determination of the in pile creep behaviour of zircaloy cladding tubes under internal and external overpressures, in FRG-2.
  Journal of Nuclear Materials 65 (1977), p. 302 - 306; also GKSS 79/E/51.
- *J. Ahlf, F.J. Schmitt*
  Bestrahlungs- und Nachuntersuchungstechnik für die Reaktordruckbehälterstahlproben bei der GKSS.
  Atomkernenergie 29 (1977), p. 128 - 132; also GKSS 77/E/25
- *F.J. Schmitt*
  Bestrahlungsuntersuchungen an einem Druckbehälterstahl für LWR.
  Zeitschrift für Werkstofftechnik, 7 (1976), p. 241 - 247; also GKSS 76/E/61
- *J. Ahlf*
  Ubersicht über die instrumentierten Bestrahlungseinrichtungen am FRG-2.
  KEST-Bericht Nr. 26 (1975), p. 9 - 23.
- *J. Ahlf, A. Reymann, H. Timm, H. Wilhelm*
  Bestrahlungseinrichtung für Korrosionsversuche unter Druckwasserreaktor-Bedingungen.
  Kerntechnik, 17 (1975), p. 135 - 139.

7.6 *INFORMATION SHEET No. 5. MAIN CHARACTERISTICS OF HOT CELL FACILITIES*

Table 3.3: Main characteristics of hot cell facilities

name of laboratory	year of active oper.	scope		gamma shielding (material, thickness, density)	leak tightness	special atmosph.	manipulators number, type (incl. spare manipulators)
		active area (m^2)	number of working positions				
old hot cell region	1963	20	1	concrete, 1,2 m; 3,3 g/cm^3	none	none	2 Mod. 8 (central research) 1 crane, 2 to.
new hot cell region (concrete cells)	1969	18	2	concrete, 1,1 m, 3,6 g/m^3	$\Delta V/V =$ 1 %/o h. ($\Delta p =$ 50 mm WS)	none	8 A 100 (Wälischmiller) 1 heavy duty manipulator 1 to. (Draht & Schrader)
		18	2	concrete, 1,1 m, 3,0 g/cm^3			
technological cells I *)	1969	7	3	lead, 10 cm	none	none	2 Mod. 7 (Hobson) several tongs
metallography cells **)	1972	10	4	lead, 17,5 cm	$\Delta V/V =$ 1 %/o h	none	7 A 200 (Wälischmiller)
chemical cells	1973	5	3	lead, 17,5 cm	$\Delta V/V =$ 1 %/o h	none	4 A 200
technological cells II ***)	1978/ 1979	8	4	lead, 20 cm	none	none	3 A 200 (Wälischmiller)

Remarks: *) impact, hardness and tensile test
**) optical microscope
***) 2500 kN tensile-test / dropweight-test

7.6.1 Owner/operator: GKSS-Forschungszentrum Geesthacht GmbH

7.6.2 Site: Geesthacht - Tesperhude (GKSS)

7.6.3 Associated reactor: FRG-1, FRG-2

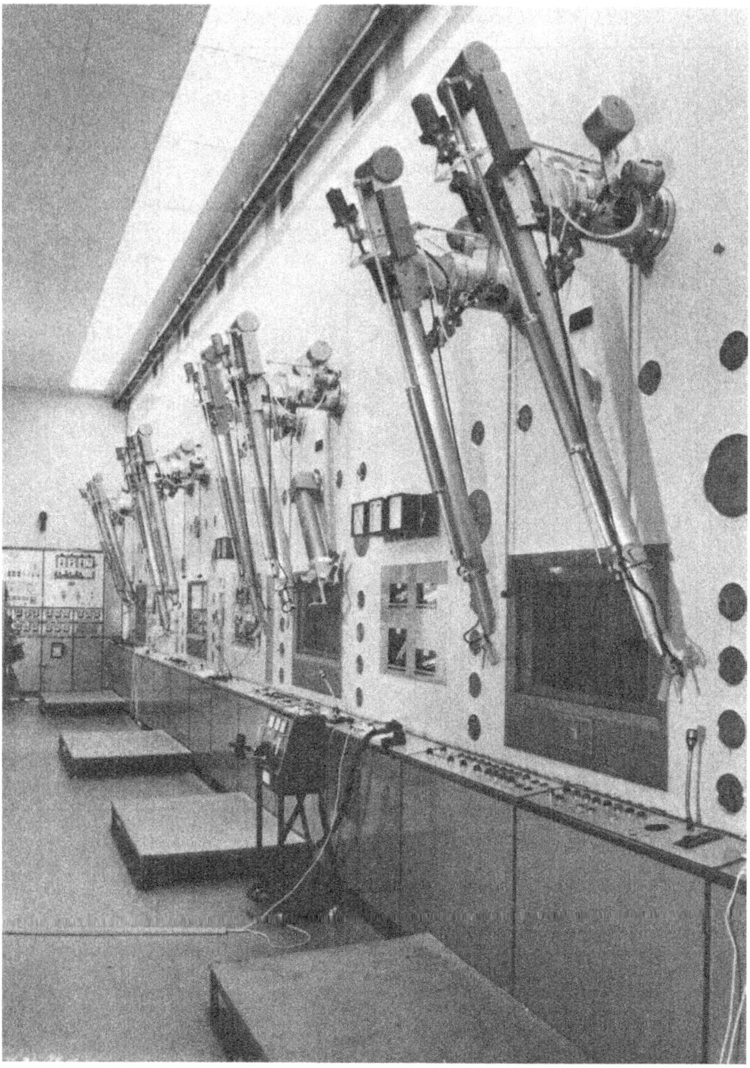

Fig. 7.5 *View of the hot cells.*

7.7 INFORMATION SHEET No. 7. EQUIPMENT AND TECHNIQUES AVAILABLE FOR POST IRRADIATION EXAMINATIONS

- *Examination of LWR fuel rods (gamma scan; burn-up determination; dimensional measurements; neutron radiography; cladding inspection by periscope, eddy current techniques and helium leak test; fission gas analysis).*
- *Examination of oxide fuel and canning material (metallography; chemical analysis; electron microprobe; stereoscan; mechanical tests).*
- *Examination of pressure vessel steels (notch impact test; drop weight test; tensile test; fracture mechanical tests).*
- *Handling of irradiation rigs.*

Relevant publications:

- *F. Borchers, W. Spalthoff and D. Wittstock*
  Das Heisse Labor der Reaktorstation Geesthacht.
  GKSS 71/E/5, presented at 13th Meeting of the Euratom-Committee for Hot Laboratories, 24-25 June, 1971 at Jülich, F.R. Germany.

7.8 INFORMATION SHEET No. 8. UTILIZATION AND SPECIALIZATION OF THE HOT CELL FACILITIES

7.8.1 *Original purposes of the facilities*

Materials research for ship reactors (LWR-type).

7.8.2 *Major modifications subsequently made*

None

7.8.3 *Major modifications planned*

None

7.8.4 *Major project and programmes for which the facilities presently are used*

- *Research programme on irradiation embrittlement of reactor pressure vessel steels (Reactor safety programme)*
- *Post irradiation examination of ship reactor fuel elements ("Otto Hahn").*
- *Research programme "Integrity of Components"*

7.8.5 *Availability of facilities for post irradiation examinations on behalf of other organizations within the European Communities*

Facilities are used for:
Examination of fuel elements and pressure vessel steel specimens from landbased reactors .

8. THE FRG 1 REACTOR AT GEESTHACHT, F.R. GERMANY. INFORMATION SHEETS.

8.1 *INFORMATION SHEET No. 1. MAIN CHARACTERISTICS OF THE REACTOR.*

Name of reactor	*Forschungs-Reaktor-Geesthacht-1 (FRG-1)*
Organization	*GKSS-Forschungszentrum Geesthacht GmbH*
Site	*Geesthacht-Tesperhude (see Fig. 7.1)*
In operation since	*1958*
Power MW (th)	*5*
Moderator	H_2O
Coolant	H_2O
Fluid and pressure in experiment channels	--
Nominal operation (days/year)	*180*
Operating cycle (days)	
on power	*5*
shut down	*2*

Responsible officer (name, address):

Dr. G. Robert

GKSS-Forschungszentrum Geesthacht GmbH
Zentralabteilung Forschungsreaktoren
D - 2054 Geesthacht

8.2 *INFORMATION SHEET No. 2 UTILIZATION AND SPECIALIZATION OF THE REACTOR*

8.2.1 *Original purposes of the reactor*

- *Beam tube experiments for fundamental and applied nuclear research, partly in collaboration with Northern German Universities.*
- *Shielding experiments for LWR.*
- *Isotope production for industry and universities requirements; Materials testing and research (organic liquids, zircaloy, steel).*

8.2.2 *Major modifications subsequently made*

Fast pneumatic tube for neutron activation analysis.
Backfitting in 1978/79.

8.2.3 *Major modifications planned*

Reduction of the enrichment in U235 in MTR-fuel elements.

8.2.4 *Major projects and programmes for which the reactor presently is used*

- *Fundamental research using beam tubes (see 8.2.1).*
- *Irradiation induced lattice defects in zircaloy and steel.*
- *Isotope production.*
- *Doping of semiconductors.*
- *Neutron activation analysis.*

8.2.5 *Loops and irradiation devices specially developed for the reactor*
see chapter 7.5.1 (FRG-2).

8.3 INFORMATION SHEET No. 3. EXPERIMENTAL FACILITIES OF THE REACTOR

8.3.1 *Irradiation positions*

Table 8.1: Irradiation positions

type/name	number	active length mm	diameter mm	typical unperturbed neutron flux density in $m^{-2}s^{-1}$ thermal	fast $(E > 0.1\ MeV)$	nuclear heating
in-core position	3	600	75	$4 - 5 \cdot 10^{17}$	$1{,}4 - 4 \cdot 10^{17}$	$6 \cdot 10^{8}$ rd/h *)
peripheral position	4	600	75	$1 - 3 \cdot 10^{17}$	$0{,}5 - 1{,}5 \cdot 10^{17}$	
pneumatic rabbit system	4	--	--	$0{,}5 - 1{,}5 \cdot 10^{17}$	$2 - 5 \cdot 10^{16}$	
beam tubes	10	--	--	$10^{16} - 10^{17}$		

*) related to Al

8.3.2 *Reactivity available for experiments:* *depending on the type of experiment.*

8.3.3 *Neutron spectra (data):* *typical neutron spectra of a light water MTR-type reactor (see also chapter 8.4).*

8.3.4 *Specialized equipment available for experimenters:*

- *pneumatic tube device for neutron activation analysis,*
- *positron-spectrometry in cloud chambers,*
- *neutron guide tubes,*
- *small angle neutron scattering,*
- *time-of-flight spectrometer,*
- *fast chopper,*
- *neutron radiography*

8.3.5	*Availability of facilities for irradiations on behalf of other organizations of the European Communities:*	*Facilities are available*
8.3.6	*Availability of hot cell facilities :*	*Facilities are available on site: see chapters 7.6 through 7.8.*
8.3.7	*Relevant publications :*	- *W. Krull, Forschungsreaktoren, GKSS 76/E/35 (1976).* - *G.A. Röbert, Der Forschungsreaktor Geesthacht als Bestrahlungsreaktor. GKSS/E/46 (1978).*

8.4 *INFORMATION SHEET No. 4. NEUTRON SPECTRA (DATA)*

See chapter 7.4

Fig. 8.1 FRG-1. View of the reactor pool.

8.5 INFORMATION SHEET No. 5. MAIN CHARACTERISTICS OF SPECIALIZED IRRADIATION DEVICES

8.5.1 *Irradiation Devices:* *see chapter 7.5.1*

8.5.2 *Relevant publications*

- *M. Nagib*
  Die Neutronenbeugungsanlage am FRG-1
  GKSS 76/E/54 (1976)
- *M. Apel et al.*
  Entwicklungsarbeiten zu einer automatisierten Messtation für die Neutronenaktivierungs-analyse.
  GKSS 76/E/50
- *F. Frisius, M. Naraghi*
  Untersuchungen von strahlengeschädigten Stahlproben mit der Neutronenkleinwinkel-streuung.
  Atomenergie 29 (1977), p. 139, and GKSS 77/E/24
- *S. Wille et al.*
  Hydrogen diffusion in metals measured by neutron scattering and absorption.
  AEM-SM-219/36 and GKSS 77/E/53.
- *F. Frisius et al.*
  Reaktordosimetrie für Bestrahlungsexperimente an der Reaktorstation Geesthacht.
  GKSS 76/E/34
- *N. Krull, P. Martens*
  Eine Einrichtung für die Bestrahlung grosser Siliziumeinkristalle.
  Kerntechnik 19 (1977), p. 131 and GKSS 77/E/26.

8.6 INFORMATION SHEET No. 6. MAIN CHARACTERISTICS OF HOT CELL FACILITIES

See chapter 7.6

8.7 INFORMATION SHEET No. 7. EQUIPMENT AND TECHNIQUES AVAILABLE FOR POST-IRRADIATION EXAMINATIONS.

See chapter 7.7

8.8 INFORMATION SHEET No. 8. UTILIZATION AND SPECIALIZATION OF THE HOT CELL FACILITIES.

See chapter 7.8

9. THE "OSIRIS" REACTOR AT SACLAY, FRANCE AND AVAILABLE HOT CELL FACILITIES. INFORMATION SHEETS.

Fig. 9.1 *Partial view of the CEN-Saclay: Avenue des Piles. In the front EL 3 (no longer in use), on the left OSIRIS-ISIS, and in the middle ORPHEE.*

9.1 INFORMATION SHEET No. 1. MAIN CHARACTERISTICS OF THE REACTOR.

Name of reactor	OSIRIS
Organization	C.E.A. - France
Site	Saclay, CEN
In operation since	1966 (Sept.)
Power MW (th)	70
Moderator	H_2O
Coolant	H_2O
Fluid and pressure in experiment channels	H_2O
Nominal operation (days/year)	230
Operating cycle (days)	
on power	28
shut down	7

Responsible officers (name, address):

OSIRIS:	*F. Cherruau*	*CEN-Saclay* *Services der Piles* *F - 91191 Gif-sur-Yvette Cedex*
Hot cells:	*M. Beyeler*	*CEN-Saclay* *Section d'Exploitation du Laboratoire d'Etudes des Combustibles Irradiés* *F - 91191 Gif-sur-Yvette Cedex*

9.2 *INFORMATION SHEET No. 2 UTILIZATION AND SPECIALIZATION OF THE REACTOR*

9.2.1 *Original purposes of the reactor*

Materials testing and isotope production

9.2.2 *Major modifications subsequently made*

- *In December 1968 the original design power of 50 MW was uprated to 70 MW.*
- *In 1979, the highly enriched uranium fuel element was changed for the low enrichment fuel "Caramel".*

9.2.3 *Major modifications planned*

None

9.2.4 *Major projects and programmes for which the reactor presently is used*

- *light water reactor development*
- *fast reactor development*
- *possibly high temperature reactor development, if asked by customers*
- *other reactor systems development*
- *safety studies*
- *radio-isotope production*

9.2.5 *Loops and irradiation devices specially developed for the reactor*

see chapter 9.5

9.2.6 *Additional information*

OSIRIS has been transformed to use low enrichment uranium (known as "Caramel" fuel) as a prototype for a possible transformation of the other research and testing reactors, in order to face a possible cut-off of highly enriched uranium supply.
The reactor has been running since the end of 1979 with the new "Caramel" fuel. It is fully operational.

Fig. 9.2 OSIRIS - Pool and core.

9.3 INFORMATION SHEET No. 3. EXPERIMENTAL FACILITIES OF THE REACTOR

9.3.1 Irradiation positions

Table 9.1: Irradiation positions

type/name	number	active length mm	useful diam. mm	typical unperturbed neutron flux density in $m^{-2}s^{-1}$ thermal	fast (E > 0,1 MeV)	nuclear heating W/g Al
central core positions *)	4 x 4	600	2 x 42	$1 \cdot 10^{18}$	$5 \cdot 10^{18}$	7
	1 x 4	600	2 x 42	$1 \cdot 10^{18}$	$3 \cdot 10^{18}$	5
peripheral core positions	7	600	84	$1{,}5 \cdot 10^{18}$	$1 \cdot 10^{18}$	3
pool locations	14	600	82	2 to $3 \cdot 10^{18}$	--	2
	12	600	82	1 to $2 \cdot 10^{18}$	--	

*) four 42 mm diameter positions can be grouped to give one 84 mm diameter position.

9.3.2 Reactivity available for experiments: not applicable

9.3.3 *Neutron spectra (data):* *see chapter 9.4*

9.3.4 *Specialized equipment available for experimenters:*
- *neutrography; gammametry*
- *fission product analysis online*
- *hydraulic rabbits for activation analysis and isotope production*
- *neutron model ISIS (700 kW)*

9.3.5 *Availability of facilities for irradiation on behalf of other organizations of the European Communities:* *Facilities are available*

9.3.6 *Availability of hot cell facilities:* *Hot cell facilities are available on the reactor site:*
- *L.E.C.I.*
- *CELIMENE and*
- *SEMIRAMIS*

(see chapter 9.6)

9.4 *INFORMATION SHEET No. 4. NEUTRON SPECTRA (DATA)*

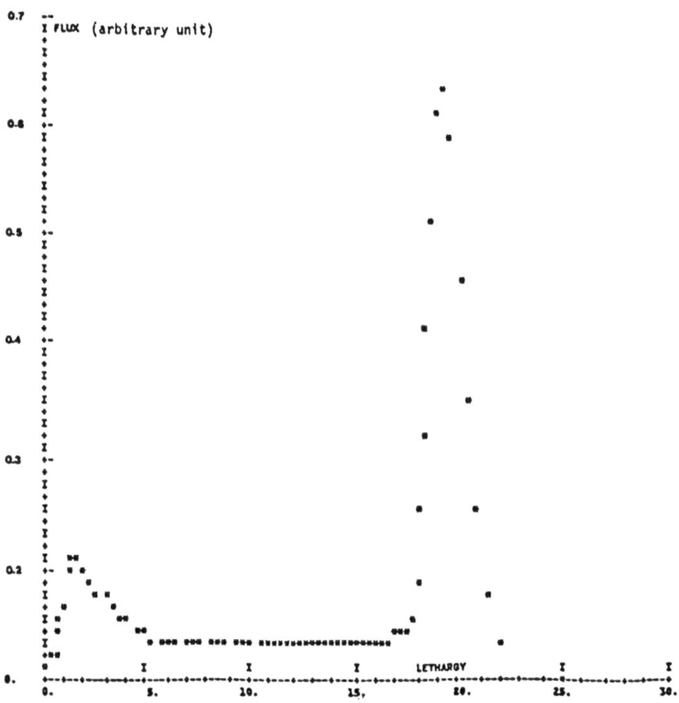

Fig. 9.3 Neutron spectrum in the reflector (3.6 cm from the tank).

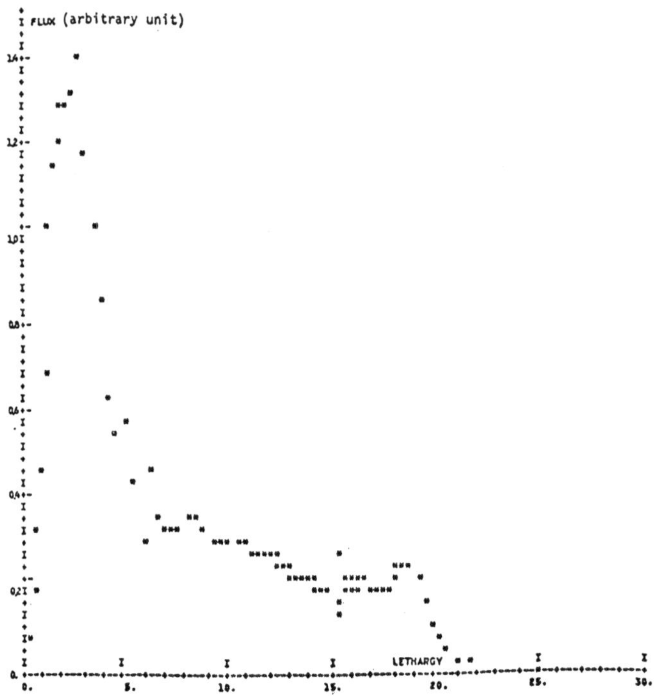

Fig. 9.4 *Neutron spectrum in in-core experiment positions (filled with steel - Ø = 28 mm - and aluminium).*

9.5 *INFORMATION SHEET No. 5. MAIN CHARACTERISTICS OF SPECIALIZED IRRADIATION DEVICES*

9.5.1 *High temperature irradiation*

- *Colibri:* *rig for irradiation in gas up to 1500 °C, fissile materials and structural materials with separate compartments and independently controlled atmosphere fission product analysis.*
- *Cythère:* *rig for continuous in-pile measurements of swelling; up to 1150°C - 0,01 mm accuracy.*
- *Florea:* *rig for continuous in-pile measurements of creep; up to 1150 °C - 0,01 mm accuracy.*
- *Spitfire:* *helium loop , 800 °C; 60 bars ; 160 g/s; 75 kW for HTR fuel rods.*

9.5.2 *Fast reactors*

- *Snac:* *sodium loop; 600 °C; 280 g/s; 1400 W/cm - 50 kW for fuel pins.*
- *Medina:* *rig for continuous in-pile measurements of cladding diameter (<5 µm) of a fuel element in NaK.*
- *Super Mammouth:* *reloadable rig ; 1000 W/cm - 25 kW, for fuel pins in NaK.*

9.5.3 Light water reactors

- *Marina:* *boiling water rig; 340 °C - 150 bars; for power cycling and safety studies on fuel pins.*
- *Boss:* *boiling water rig; reloadable in hot cells for re-irradiation of short fuel pins.*
- *Cadence:* *boiling water rig, reloadable in the pool of the reactor, for re-irradiation of long fuel pins (up to 1,80 m).*
- *Colibri:* *rig H_2O - 170 bars for dynamic corrosion studies.*
- *Colibri:* *rig NaK - 300 °C - cladding creep*
- *Nicobra-Mirette:* *rig for structural materials irradiation and sensors in gas medium - 300 - 600 °C.*
- *Siat:* *rig for pressure vessel steel samples irradiation in gas medium - 300 °C - large capacity.*
- *Isabelle:* *water loop ; 300 °C - 150 bars - 3,5 m/s - 500 W/cm for cycling of one fuel pin; operating with high activity - possibility for four pins.*
- *Irène:* *water loop; 300 °C - 155 bars - 6 m/s - 140 kW - 700 W/cm for power cycling of 4 or 8 UO_2 rods; reloadable; with fission products analysis in line.*

9.6 INFORMATION SHEET No. 6. MAIN CHARACTERISTICS OF HOT CELL FACILITIES

9.6.1 The characteristics of the hot cell facilities are summarized in table 9.2.

9.6.2 Owner / operator: *The L.E.C.I. hot laboratory / Département de Technologie, Section d'Exploitation du L.E.C.I.*

9.6.3 Site: *CEA-CEN Saclay*

9.6.4 Associated reactor: *These facilities are not especially associated with a reactor.*

9.7 INFORMATION SHEET No. 7. EQUIPMENT AND TECHNIQUES AVAILABLE FOR POST IRRADIATION EXAMINATION

9.7.1 L.E.C.I.

for fuel rods up to 4.000 mm length:

- *visual examination*
- *photography*
- *metrology*
- *γ-spectrometry*
- *radiography and micro-radiography*
- *neutron radiography*
- *leak tests*
- *cladding check by Eddy current*
- *dismantling*
- *cleaning (Na and NaK neutralisation)*
- *machining (cutter, lathe, milling-machine, drill, core-sampler, etc.)*
- *argon-arc welding*
- *chemical dissolution*

Table 9.2: Main characteristics of hot cell facilities

name of laboratory	year of active operation	scope: active area (m^2)	scope: number of working pos.	gamma shielding (material, thickness, and density)	leak tightness	special atmosphere	manipulators, number and type
L.E.C.I	1959 extension 1964 and 1970	1000 (hot cell area 80)	23	<u>K.line</u> 5 cells : concrete, 1,03 m, 3.5 g/cm^3 6 cells: concrete, 1,03 m, 2,4 g/cm^3 2 cells : lead, 0,2 m 1 cell: iron, 0,5 m <u>I.line</u> 3 cells : lead in iron container, 0,27 m 3 cells : lead in iron container, 0,17 m	β, γ	air	24 - CRL M8 12 - CRL compact (the large cells are equipped with special XYZ cranes, 400 kg 150 kg and 100 kg)
CELIMENE	1964 (modification 1975)	200 (hot cell area 80)	2	concrete, 1,03 m, 3,5 g/cm^3	β, γ	air	4 CRL ME (heavy manipulators 300 kg; 2 cranes, 1000 kg, 200 kg resp.; 1 elevator)
SEMIRAMIS	1971	50 (5)	4	lead, 0,15 m	α, β, γ	air	4 CRL MG 2 CRL M7

for samples:

- *o mechanical tests*
- *o weighing*
- *o density measurement*
- *o internal cladding temperature m easurement (85 Kr)*
- *o recovery and analysis of fission and other nuclear reaction gases*
- *o burn-up measurement*
- *o thermal conductivity measurement*
- *o radiocrystallography*
- *o metallography (conventional, ionic attack)*
- *o autoradiography and alphagraphy*
- *o optical microscope examination*
- *o shielded microprobe examination*

9.7.2 CELIMENE

for PWR fuel assembly:

- *o metrology*
- *o visual examination (T.V. périscope)*
- *o dismantling machine*
- *o devices for fuel rods extraction*

9.8 INFORMATION SHEET No. 8. UTILIZATION AND SPECIALIZATION OF THE FACILITIES

9.8.1 Original purposes of the facilities

- *The hot cell CELIMENE was specially designed to provide technical support for the EL 3 experimental reactor.*
- *The hot cells of the LECI facility were operated on the UNGG programme.*

9.8.2 Major modifications subsequently made

- *The hot cell CELIMENE has been converted in 1975 to the examinations of LWR power fuel assemblies (examination - metrology - dismantling and machining)*
- *The LECI laboratory has been converted to the examination,metrology and special examination on PWR fuel rods of 4.000mm length.*
  Large development of gamma spectrometry.

9.8.3 Major modifications planned

It is planned to equip the facilities for examinations of the 1.300 MW_{el} PWR fuel assemblies and fuel rods.

9.8.4 Major projects and programmes for which the facilities presently are used

- *PWR (fuel assemblies, fuel rods, structural materials)*
- *FBR (structural materials)*

9.8.5 Availability of facilities for post irradiation examinations on behalf of other organizations within the European Communities

Facilities are available

10. THE "ORPHEE" REACTOR AT SACLAY, FRANCE. INFORMATION SHEETS.

10.1 *INFORMATION SHEET No. 1. MAIN CHARACTERISTICS OF THE REACTOR.*

Name of reactor	*ORPHEE*
Organization	*I.R.F. - C.E.A. France*
Site	*Saclay, C.E.N. (see Fig. 9.1)*
In operation since	*1980*
Power MW (th)	*14*
Moderator	D_2O - H_2O
Coolant	H_2O
Fluid and pressure in experiment channels	
Nominal operation (days/year)	*250*
Operating cycle (days)	
on power	>100
shut down	*7*

Responsible officer (name, address):

F. Cherruau — *CEN-Saclay*
Services des Piles
F - 91191 Gif-sur-Yvette Cedex

10.2 *INFORMATION SHEET No. 2 UTILIZATION AND SPECIALIZATION OF THE REACTOR*

10.2.1 *Original purposes of the reactor*

To provide an intense source of neutron beams for neutron diffraction and inelastic neutron scattering applied to fundamental research on condensed matter.

10.2.2 *Major modifications subsequently made*

not applicable

10.2.3 *Major modifications planned*

not applicable

Fig. 10.1 ORPHEE reactor. Reactor block with beam-holes' exits.

10.2.4 *Major projects and programmes for which the reactor presently is used*

- *Fundamental research on condensed matter*
- *Activation analysis*
- *Radioisotope production*
- *Silicon irradiation*
- *Industrial neutron radiography*

10.2.5 *Loops and irradiation devices specially developed for the reactor*

not applicable

10.2.6 *Additional information*

ORPHEE is a pool-type research reactor specially designed for generating high intensity neutron beams. Its thermal power is approximately 14 MW.
The compact-design core uses light water as coolant; to allow for production of high intensity thermal neutron beams the reflector of the reactor is a heavy water reflector where the thermal flux reaches $3 \cdot 10^{18}$ *n* $m^{-2}s^{-1}$.
In addition some vertical channels are devoted to irradiations.

10.3 INFORMATION SHEET No. 3. EXPERIMENTAL FACILITIES OF THE REACTOR

10.3.1 *Irradiation positions*

Table 10.1: Irradiation positions

type/name	number	active length mm	useful diam. mm	typical unperturbed neutron flux density in $m^{-2}s^{-1}$ thermal	fast (E > 0,1 MeV)	nuclear heating W/g Al	remarks
I 1	1	900	110	$2 \cdot 10^{18}$	**)	**)	
I 2	1	900	110	$3 \cdot 10^{18}$			radioisotope production
I 3	1	900	110	$3 \cdot 10^{18}$			
R 1, R 2	2	900	154	$3 \cdot 10^{17}$			silicon irradiation
P 1, P 2	2	180 *)	36 *)	$2{,}5 \cdot 10^{17}$			
P 3	1	57 *)	20 *)	$2{,}5 \cdot 10^{17}$			pneumatic rabbits
P 4	1	57 *)	20 *)	$4 \cdot 10^{17}$			
horizontal channels	9	**)	**)	$3 \cdot 10^{18}$			20 beams: 8: thermal neutrons 8: cold neutrons 4: hot neutrons

*) sample dimensions; **) not applicable

10.3.2 *Reactivity available for experiments:* *not applicable*

10.3.3 *Neutron spectra (data):* *not yet available.*

10.3.4	*Specialized equipment available for experimenters:*	- *Industrial neutron radiography* - *Hot source for beam experiments (short wave-length neutrons)* - *Cold source for beam experiments (long wave-length neutrons)* - *6 neutron guides*
10.3.5	*Availability of facilities for irradiation on behalf of other organizations of the European Communities:*	*Facilities are available*
10.3.6	*Availability of hot cell facilities:*	*Hot cell facilities are available on the reactor site (see chapter 9.6).*
10.3.7	*Relevant publications:*	*not yet applicable*

10.4 INFORMATION SHEET No. 4. NEUTRON SPECTRA (DATA)

Spectra not yet avialable

10.5 INFORMATION SHEET No. 5. MAIN CHARACTERISTICS OF SPECIALIZED IRRADIATION DEVICES

Not applicable

10.6 INFORMATION SHEET No. 6. MAIN CHARACTERISTICS OF HOT CELL FACILITIES

See chapter 9.6

10.7 INFORMATION SHEET No. 7. EQUIPMENT AND TECHNIQUES AVAILABLE FOR POST IRRADIATION EXAMINATION

See chapter 9.7

10.8 INFORMATION SHEET No. 8. UTILIZATION AND SPECIALIZATION OF THE HOT CELL FACILITIES

See chapter 9.8

11. THE "SILOE" REACTOR AT GRENOBLE, FRANCE AND ASSOCIATED HOT CELL FACILITIES. INFORMATION SHEETS.

Fig. 11.1 *View of the SILOE reactor.*

11.1 *INFORMATION SHEET No. 1. MAIN CHARACTERISTICS OF THE REACTOR.*

Name of reactor	*SILOE*
Organization	*C.E.A. France*
Site	*Grenoble, CEN*
In operation since	*1963*
Power MW (th)	*35*
Moderator	H_2O
Coolant	H_2O
Fluid and pressure in experiment channels	
Nominal operation (days/year)	*235*
Operating cycle (days)	
on power	*21*
shut down	*7*

Responsible officer (name, address):

F. Merchie	*CEN-Grenoble* *85 X* *F - 38041 Grenoble Cèdex*

11.2 INFORMATION SHEET No. 2 UTILIZATION AND SPECIALIZATION OF THE REACTOR

11.2.1 Original purposes of the reactor

- Materials testing,
- Basic research with neutron beams,
- Radioisotopes production,
- Activation analysis.

11.2.2 Major modifications subsequently made

- In 1968 the original design power of 15 MW was uprated to 30 MW
- In 1972 the nominal power level was uprated to 35 MW.
- In 1977 the replacement of core support was made in order to irradiate or re-irradiate fuel pins of active length up to 1,60m.

11.2.3 Major modifications planned

None

11.2.4 Major projects and programmes for which the reactor presently is used

- Fast reactor development
- High temperature reactor development
- Light water reactor development (pressurized and boiling water)
- Gas-cooled reactor development
- Neutron spectrometry
- Radioisotopes production

11.2.5 Loops and irradiation devices specially developed for the reactor

The irradiation devices are standardized. Adaptations to other testing reactors have been made when necessary.
Main characteristics of specialized irradiation devices are listed in chapter 11.5.

11.3 INFORMATION SHEET No. 3. EXPERIMENTAL FACILITIES OF THE REACTOR

11.3.1 Irradiation positions

Table 11.1: Irradiation positions

type/name	number	active length mm	useful diam. mm	typical unperturbed neutron flux density in $m^{-2}s^{-1}$ thermal	fast ($E >$ 0,1 MeV)	nuclear heating W/g Al
core positions	12	600	36	$4 \cdot 10^{18}$	$5 \cdot 10^{18}$	12
peripheral positions	40	600	77	$5 \cdot 10^{18}$	$2 \cdot 10^{18}$	7
beam tubes	2	--	–	$4 \cdot 10^{13}$ *)		

*) at the outer end of a standard collimator

Fig. 11.2 General view of SILOE swimming pool.

11.3.2 Reactivity available for experiments: *2 °/o Δk/k*

11.3.3 Neutron spectra (data): *see chapter 11.4*

11.3.4 Specialized equipment available for experimenters:

Neutrography, on line computers, in-pile gamma spectrometry, gaseous fission product sampling and analysis during irradiation, dismantling hot cell.

11.3.5 Availability of facilities for irradiation on behalf of other organizations of the European Communities:

Facilities are available. For the moment the facilities are used by CNEN (Italy), GfK Karlsruhe and KFA Jülich (F.R. Germany), UKAEA and CEGB (U.K.), Joint Research Centre (Karlsruhe), and IRE (Belgium).

11.3.6 Availability of hot cell facilities:

Hot cell facilities are available on the reactor site:
- Dismantling HOT CELL in the reactor hall
- LAMA-DMG
For details see chapter 11.6

11.4 INFORMATION SHEET No. 4. NEUTRON SPECTRA (DATA)

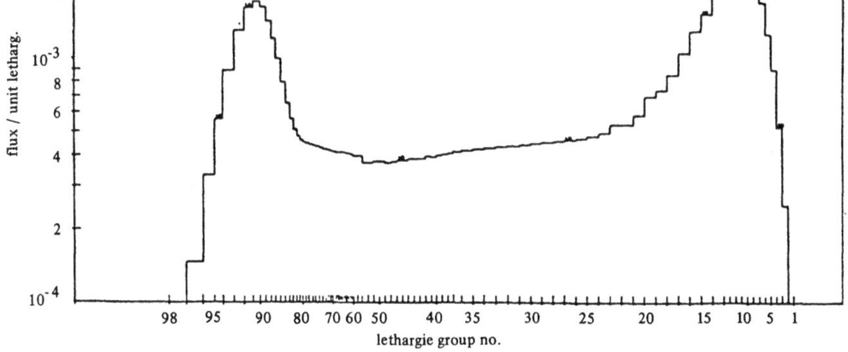

Fig. 11.3 SILOE-reactor. Typical neutron spectra.

11.5 *INFORMATION SHEET NO. 5. MAIN CHARACTERISTICS OF SPECIALIZED IRRADIATION DEVICES*

- *CHOUCA* — *rig for structural and fissile materials irradiation, 200 - 900 °C, with 5 different types of sample holders for graphite, dispersed fuel, creep, swelling, growth and corrosion studies.*

- *HEBE* — *rig for structural materials irradiation, 350 - 1400 °C, 60 bars.*

- *CYRANO* — *rig for fuel pin irradiation (in Na or NaK, 300 - 800 °C, from 1 to 155 bars) with continuous measurement of the the fuel sample nuclear power.*
  Sample diameter 4 - 35 mm.
  This rig was used for fission product studies (release, pressure, diffusion) and for dimensional measurements (length, diameter and profile).

- *HF* — *High Frequency Furnaces for small fissile material samples up to 2000 °C (high burn-up irradiation).*

- *Thermopump sodium loop* — *for experimental irradiation of fuel pins (up to 3), 70 kW, 2 - 6 m/s flow speed, temperature regulation. Cladding failure studies through fission products release and by gamma scanning during irradiation.*

- *BOUFFON, AQUILON and GRIFFON* — *rigs for fuel pins irradiations of the light water reactor family with possibility of water sampling (water chemistry, fission product recuperation, gamma spectrometry) cladding failure evolution studies and profiles and diameters measurements.*
  These rigs are used for irradiation or re-irradiation of fuel rods of Nuclear Power Plant.

- *CO_2 loop* — *used to study graphite corrosion under irradiation.*

- *COMEDIE loop* — *pressurized helium loop (60 bars), 50 g/s, for high-temperature fuel irradiations (900 °C gas outlet) to study fuel evolution, fuel products migration and deposition under various chemical environments.*

- *TUMULT* — *rig with separate internal divisions for irradiation of high temperature fuel, 800 - 1500 °C. Possibility of thermal cycling. Measurement of the fission product release rate. Gamma spectrometry.*

- *Resonant cavity* — *methods of accurate in-pile measurement of dimensional changes can be adapted to HEBE, CHOUCA, HF Furnace, Thermopump.*

- *JET-PUMP loop* — *pressurized water loop to do fuel pin irradiation in PWR conditions (320 °C - 150 bars - 4 m/s). This rig is connected to a water feeding system (fission-product degassing).*

- *On-line chemical measurements* — *Possibilities of power cycling and corrosion experiment.*

Table 11.2: Main characteristics of hot cell facilities

name of laboratory	year of active operation	scope: active area (m^2)	scope: number of working pos.	gamma shielding (material, thickness, and density)	leak tightness	special atmosphere	manipulators, number and type in brackets: main use of cells
Siloe reactor Hot Cell	1963	9	1	heavy concrete 1 m, 3,5 g/cm^3	–	negative atmospheric pressure inside	2 PYE; CD 3 985 (equipped with various special tools used for dismantling, loading and reloading of rigs, capsules, etc.)
LAMA/DMG (laboratory for the analysis and measurement of high-level radioactivity)	1961	14	3	barytic concrete, 1 m 3,5 g/cm^3	–	–	6 PYE model 8 (machining)
	1962	10	9	lead, 0,15 m, 10,5 g/cm^3	–	–	14 Hobson model 7 (metallographic tests)
	1965	19	3	barytic concrete, 1 m 3,5 g/cm^3	air exchange	nil –	6 PYE model 8 (sectioning X-ray test)
	1966	8	5	lead, 0,20 m	air exchange 10 - 50 times per h.	–	10 CRL model 9 (non-destructive testing)
	1966	17	12	lead, 0,15 m	15 mm WG negative pressure	–	swivel-jointed tongs (strength of materials)
	1970	7	6	lead, 0,15 m	15 mm WG negative pressure	–	10 Hobson, model 7 (HTR-specific tests)
	1973	6	5	lead, 0,15 m	15 mm WG negative pressure	–	10 Hobson, model 7 (metallographic tests)

11.6 INFORMATION SHEET No. 6. MAIN CHARACTERISTICS OF HOT CELL FACILITIES.

11.6.1 A summary of the data on the available hot cell facilities has been given in Table 11.2.

11.6.2 Owner / operator — CEA-CEN Grenoble

11.6.3 Site — Grenoble, CEN

11.6.4 Associated reactors — MELUSINE, SILOE and, if required, other CEA reactors.

11.7 INFORMATION SHEET No. 7. EQUIPMENT AND TECHNIQUES AVAILABLE FOR POST-IRRADIATION EXAMINATIONS

- Stripping of all testpieces irradiated in research pile - various machining operations (sectioning, unselting, removal of samples and testpieces by spark erosion).
- Non-destructive testing: observation, viewing through periscopes and macroscopes X-ray photography, high-accuracy metrology (diameters, excentricity, bowing, length), Eddy current, density measurements, γ-spectrometry and γ-scanning.
- Destructive testing: fission gases puncturing, testing of material strength (hardness, tensile strength, toughness - Charpy, Pellini tests, etc.
- Metallography: sample preparation and examination by optical microscopy, Preparation of samples for examination by transmission and scanning, electron microscopy or microprobe.

11.8 INFORMATION SHEET No. 8. UTILIZATION AND SPECIALIZATION OF THE HOT CELL FACILITIES

The Laboratory for the Analysis and Measurement of high-level Activity (LAMA) was conceived in 1958 to meet the needs of the newly commissioned MELUSINE research reactor. The building was extended during 1964 - 65 in order to fulfil the new requirements linked with the criticality of SILOE.
This Laboratory has never had a specific task related to a clearly defined research programme, on the contrary, the highly diverse nature of its equipment enables it to adapt easily to the wide range of testpieces irradiated in the piles at Grenoble.

The additional capacity installed in the LAMA since 1965 meets two requirements, namely
- for specific studies concerning new families of reactors (e.g. installation of cells for HTR particles)
- for an increase in the LAMA's potential in a field already exploited (metallographic testing, for example).

The non-specialist nature of the Laboratory enables it to offer its services for post-irradiation tests in connection with all reactor families and for all public or private bodies in the European Communities. Its exploitation by several European organizations has proven this in the past.

12. THE "MELUSINE" REACTOR AT GRENOBLE, FRANCE, AND ASSOCIATED HOT CELL FACILITIES. INFORMATION SHEETS.

12.1 *INFORMATION SHEET No. 1. MAIN CHARACTERISTICS OF THE REACTOR.*

Name of reactor	*MELUSINE*
Organization	*C.E.A. - France*
Site	*Grenoble, CEN (see Fig. 11.1)*
In operation since	*1958*
Power MW (th)	*8*
Moderator	H_2O
Coolant	H_2O
Fluid and pressure in experiment channels	
Nominal operation (days/year)	*260*
Operating cycle (days)	
on power	*23*
shut down	*5*

Responsible officer (name, address):

F. Merchie — *CEN-Grenoble*
85 X
F-38041 Grenoble Cédex

12.2 *INFORMATION SHEET No. 2 UTILIZATION AND SPECIALIZATION OF THE REACTOR*

12.2.1 *Original purposes of the reactor*

- *Testing of structural materials and fuel samples in instrumented capsules.*
- *Solid state physics at very low temperature and under magnetic field.*
- *Nuclear physics on beam tubes (neutron diffraction, Mössbauer effect application, short - lived fission products, etc.)*
- *Radioactivation analysis.*

12.2.2 *Major modifications subsequently made*

The power was uprated several times; in 1971 a power of 8 MW was reached (Applied and fundamental research programme)

12.2.3 *Major modifications planned*

None

Fig. 12.1 General view of MELUSINE reactor block.

12.2.4 *Major projects and programmes for which the reactor presently is used*

- *Reactor materials development, for light water reactors, fast reactors and high temperature reactors.*
- *Radio isotopes production*
- *Neutron spectrometry*
- *Nuclear physics and chemistry*

12.2.5 *Loops and irradiation devices specially developed for the reactor*

- *CHOUCA - rigs for structural or fissile materials 200 - 900 °C*
- *HEBE - rigs for structural materials, 350 - 1400 °C*
- *CYRANO - rigs for fissile materials in NaK at 900 °C*
- *High Frequency Furnace (2000 °C)*
- *Cold loops - liquid helium, liquid neon, liquid nitrogen*

For main characteristics of these irradiation devices, *see chapter 11.5*

12.3 INFORMATION SHEET No. 3. EXPERIMENTAL FACILITIES OF THE REACTOR

12.3.1 *Irradiation positions*

Table 12.1: Irradiation positions

type/name	number	active length mm	useful diam. mm	typical unperturbed neutron flux density in $m^{-2}s^{-1}$ thermal	fast (E > 0,1 MeV)	nuclear heating W/g Al
core positions	4	600	36	$7 \cdot 10^{17}$	$1 \cdot 10^{18}$	3
peripheral positions	30	600	77	$9 \cdot 10^{17}$	$5 \cdot 10^{17}$	1,5
beam tubes	5	--	--	$7{,}8 \cdot 10^{12}$ *)	--	--

*) *At the outer end of a standard collimator. The thermal flux at the neutron source surface of the beam tube is* $9 \cdot 10^{17}\ cm^{-2}\ s^{-1}$

12.3.2 *Reactivity available for experiments:* *2 °/o Δk/k*

12.3.3 *Neutron spectra (data):* *see chapter 12.4*

12.3.4 *Specialized equipment available for experimenters:*

- *neutron radiography*
- *hot cell*
- *automatic data logging*
- *sampling and analysis of gaseous fission products*
- *cold loops: liquid nitrogen, liquid neon, liquid helium*
- *3 pneumatic rabbits: 1 in* D_2O*, 1 fast rabbit in* D_2O*, 1 transfer rabbit*
- D_2O *tank with various irradiation positions.*

12.3.5 *Availability of facilities for irradiation on behalf of other organizations of the European Communities:* *Facilities are available*

12.3.6 *Availability of hot cell facilities:*

Hot cell facilities are available on the reactor site:

- *Dismantling cell in the reactor hall*
- *LAMA - DMG*

For details see chapter 11.6

12.3.7 *Relevant publications:*

- *CEA/B.I.S.T. No. 156 (Febr. 1971)*
- *CEA/B.I.S.T. No. 153 and 154 (Nov.-Dec. 1970)*

12.4 INFORMATION SHEET No. 4. NEUTRON SPECTRA (DATA)

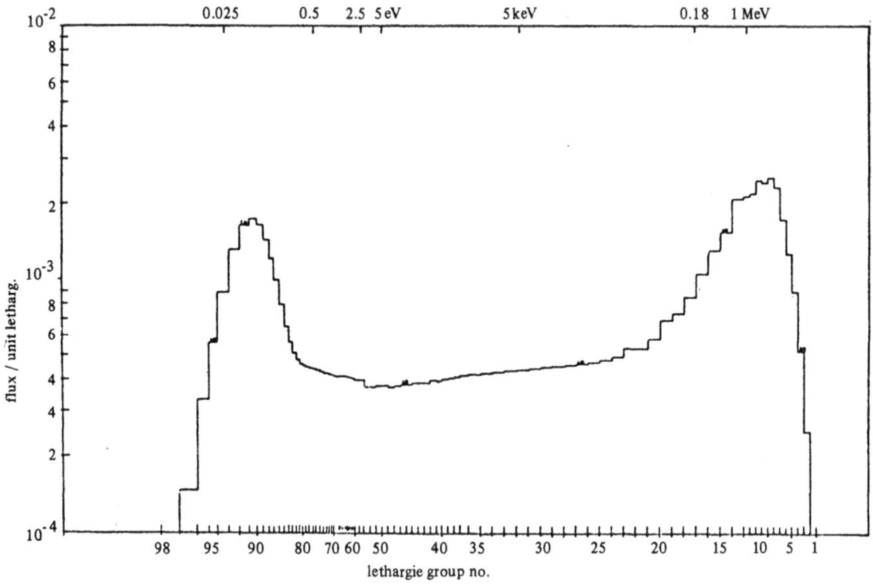

Fig. 12.2

Fig. 12.2 MELUSINE reactor. Typical neutron spectra (same as Fig. 11.3).

12.5 INFORMATION SHEET No. 5. MAIN CHARACTERISTICS OF SPECIALIZED IRRADIATION DEVICES

See chapter 11.5

For relevant publications see chapter 12.3.7.

12.6 INFORMATION SHEET No. 6. MAIN CHARACTERISTICS OF HOT CELL FACILITIES

Table 12.2 : Main characteristics of hot cell facilities

name of laboratory	year of active oper.	scope		gamma shielding (material, thickness, density)	leak tightness	special atmosph	manipulators number, type (incl. spare manipulators)
		active area (m^2)	number of working positions				
MELUSINE Hot Cell	1960	4	1	heavy concrete 1 m, 3,2 g/cm^3	--	negative pressure	1 manipulator PYE
LAMA-DMG	see chapter 11.6						

12.7 INFORMATION SHEET No. 7. EQUIPMENT AND TECHNIQUES AVAILABLE FOR POST-IRRADIATION EXAMINATION

See chapter 11.7

12.8 INFORMATION SHEET No. 8. UTILIZATION AND SPECIALIZATION OF THE HOT CELL FACILITIES

See chapter 11.8

13. THE HIGH FLUX REACTOR AT GRENOBLE, FRANCE. INFORMATION SHEETS.

Fig. 13.1 The High Flux Reactor building at Grenoble.

13.1 INFORMATION SHEET No. 1. MAIN CHARACTERISTICS OF THE REACTOR.

Name of reactor	*High Flux Reactor*
Organization	*Max von Laue - Paul Langevin Institute*
Site	*Grenoble, France*
In operation since	*1971*
Power MW (th)	*57*
Moderator	D_2O
Coolant	D_2O
Fluid and pressure in experiment channels	--
Nominal operation (days/year)	*264*
Operating cycle (days)	*56*
on power	*44*
shut down	*12*

Responsible officer (name, address) :

F. Franzetti *Institut Max von Laue - Paul Langevin*
156 X
38042 GRENOBLE Cedex
France

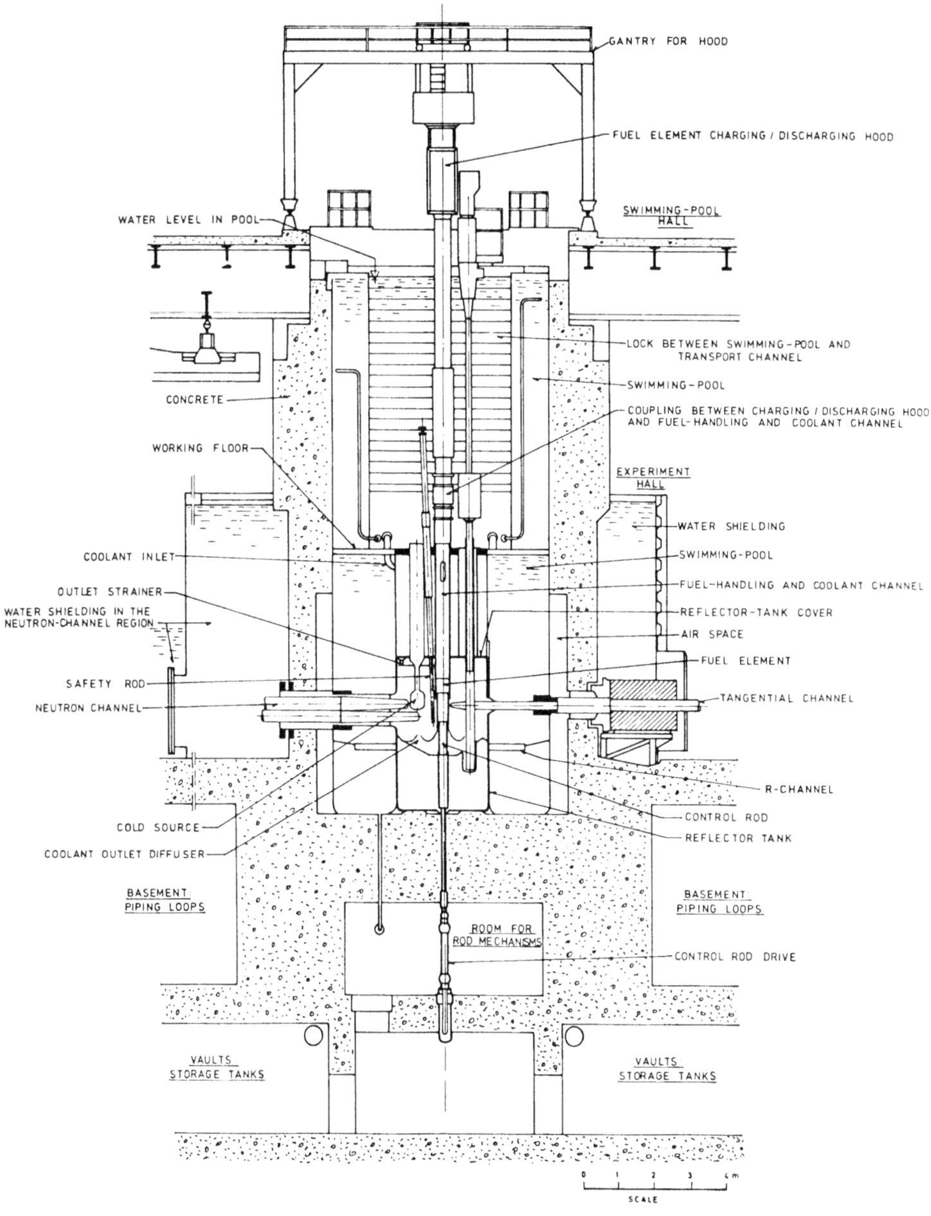

Fig. 13.2 Vertical cross-section of the Grenoble High Flux Reactor.

13.2 INFORMATION SHEET No. 2 UTILIZATION AND SPECIALIZATION OF THE REACTOR.

13.2.1 Original purposes of the reactor

Beam reactor

13.2.2 Major modifications subsequently made

Several modifications of existing instruments in order to develop new fields of research.

13.2.3 Major modifications planned

New Cold Source with vertical ultra cold neutron beam tube.

13.2.4 Major projects and programmes for which the reactor presently is used

fundamental research

13.2.5 Loops and irradiation devices specially developed for the reactor

not applicable

13.3 INFORMATION SHEET No. 3. EXPERIMENTAL FACILITIES OF THE REACTOR

13.3.1 Experimental facilities

Table 13.1 : Experimental facilities

type/name	number	length (cm)	thermal flux at nose of beam tube, in $m^{-2}s^{-1}$	remarks
horizontal beam tubes	9	~ 450	$5 - 8 . 10^{18}$	
vertical beam tubes	4	~ 500	$5 - 8 . 10^{18}$	
going through beam tube	1		$\sim 6 . 10^{18}$	
vertical irradiation tubes	2	~ 500	$1{,}3 . 10^{19}$	removed
short time irradiation : a) in D_2O	1	~ 500	$3 . 10^{18}$	max. 2 h irradiation removed
b) in H_2O	2	~ 500	$3 . 10^{17}$	max. 4 h irradiation 1 tube removed
liquid D_2, Cold Source	1			~ 25 K for 1 inclined tube and 5 neutron guides
graphite, Hot Source	1			~ 2 400 K for 3 vertical beam tubes

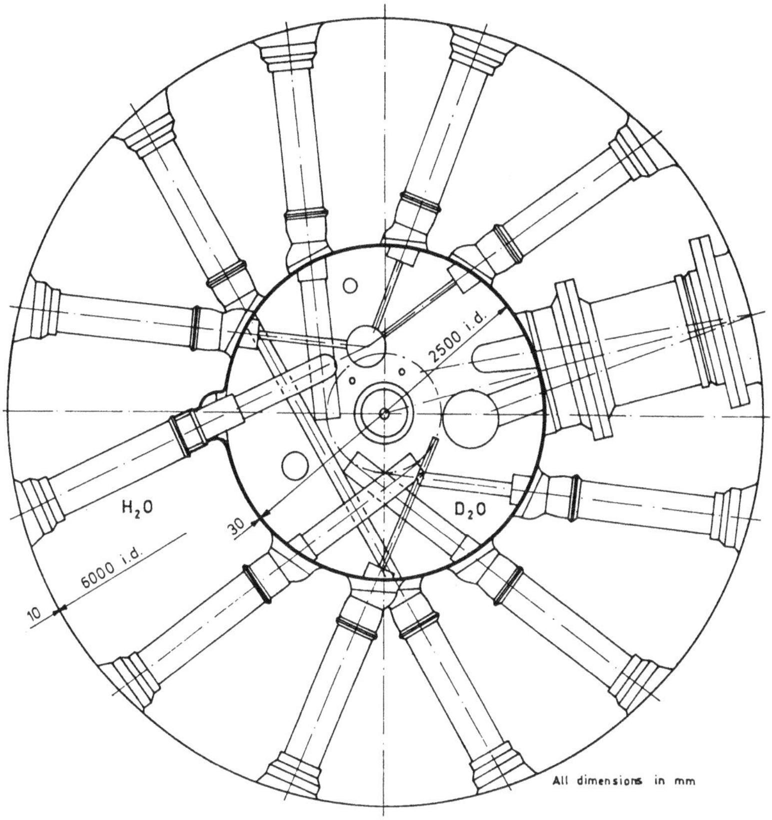

Fig. 13.3 Arrangement of experimental facilities in the High Flux Reactor at Grenoble.

13.3.2	*Reactivity available for experiments:*	*5 %*
13.3.3	*Neutron spectra (data):*	*see chapter 13.4*
13.3.4	*Specialized equipment available for experimenters:*	*about 36 instruments for fundamental research*
13.3.5	*Availability of facilities for irradiation on behalf of other organizations of the European Communities:*	*Facilities are not available*
13.3.6	*Availability of hot cell facilities:*	*not available*
13.3.7	*Relevant publications:*	*Annual Reports ILL*

13.4 INFORMATION SHEET No. 4. NEUTRON SPECTRA (DATA)

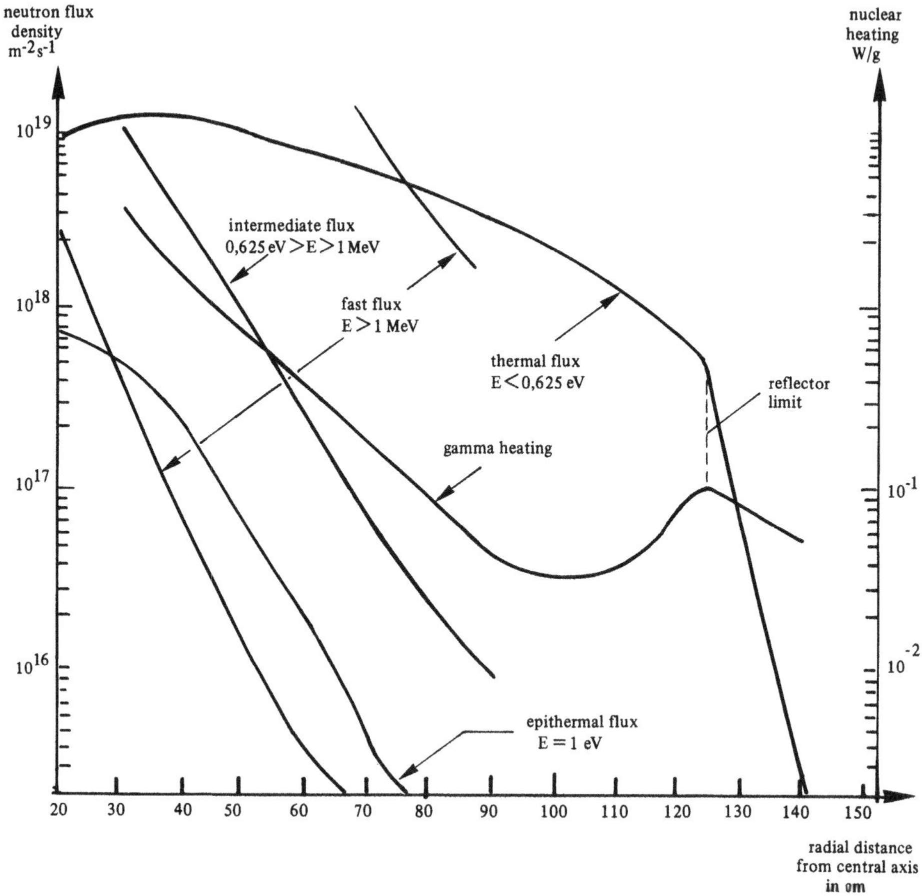

Fig. 13.4 Distribution of neutron flux density and nuclear heating in the horizontal core midplane.

13.5 INFORMATION SHEET No. 5. MAIN CHARACTERISTICS OF SPECIALIZED RESEARCH FACILITIES

Detailed information is available at ILL

14. THE "TRITON" REACTOR AT FONTENAY-AUX-ROSES, FRANCE AND ITS ASSOCIATED HOT CELL FACILITIES. INFORMATION SHEETS.

14.1 *INFORMATION SHEET No. 1. MAIN CHARACTERISTICS OF THE REACTOR.*

Name of reactor	*TRITON*
Organization	*C.E.A. - France*
Site	*Fontenay-aux-Roses, CEN*
In operation since	*1959*
Power MW (th)	*6,5 (since 1965)*
Moderator	H_2O
Coolant	H_2O
Fluid and pressure (kg/cm²) *experiment channels*	H_2O
Nominal operation (days/year)	*250*
Operating cycle (days)	
on power	*32*
shut down	*7*

Responsible officers (name, address):

TRITON	*F. Cherruau*	*CEN - Saclay* *Services des Piles* *F 91191 Gif-sur-Yvette Cedex*

14.2 *INFORMATION SHEET No. 2 UTILIZATION AND SPECIALIZATION OF THE REACTOR*

14.2.1 *Original purposes of the reactor*

Since it was mainly intended for studies on protection against radiation, the reactor core could be moved for irradiation purposes in various parts of the pool. Operation was with natural or forced draught for a given setup, with a maximum power of 1,2 MW.

14.2.2 *Major modifications subsequently made*

The pile was completely modified for the purpose of specializing in irradiation in fixed positions and of increasing the number of irradiation positions.
Following this modification the operating power of TRITON increased to 6,5 MW.
In view of this new potential, experimental vues have been directed towards the irradiation of fuel, graphite, cladding and structural materials, protective devices, radio isotope production, pure research.

14.2.3 *Major modifications planned*

Final shut-down in 1981.

14.2.4 *Major projects and programmes for which the reactor presently is used*

Currently the reactor is being used :

- *for an extensive programme covering the irradiation of low gamma heat-up reactor vessel steels,*
- *to complete the thermoelectronic studies,*
- *to supply radio isotopes at all times, thanks to a reactor availability of over 8000 hours/year.*

This availability also permits the further development of Industrial Neutron Radiography in high efficiency plant which is unique in France and even in Europe.

Finally the reactor is used for pure research, basically on low-temperature studies and on the study of metal dispersion.

14.2.5 *Loops and irradiation devices specially developed for the reactor*

- *A type of loop developed for this reactor has since been used in other CEA reactors, namely low-temperature loops incorporating a cryogenerator.*
- *Mention should be made of the high-performance "FRUIT"-type irradiation devices and also of conventional devices: Furnaces, CHOUCA, HEBE, COLIBRI, NICOBRA.*

These simple inexpensive devices have been developed for the irradiation of protective materials (concretes) and vessel steels ("BETSY" and "SIAT").

14.3 INFORMATION SHEET No. 3. EXPERIMENTAL FACILITIES OF THE REACTOR

14.3.1 Irradiation positions

Table 14.1 : Irradiation positions

type/name	number	active length mm	useful diam. mm	typical unperturbed neutron flux density in $m^{-2}s^{-1}$ thermal	fast ($E > 0,1\,MeV$)	nuclear heating W/g Al	remarks
in core 25	1	600	$\leqslant 70$	$\leqslant 4,2 . 10^{17}$	$\leqslant 3,5 . 10^{17}$	$\leqslant 1,7$	five positions in first periphery with lead shield
44	1	600	$\leqslant 70$	$\leqslant 4,2 . 10^{17}$	$\leqslant 3,5 . 10^{17}$	$\leqslant 1,7$	
first periphery	17	600	$\leqslant 70$	$2 - 6 . 10^{17}$	$0,3 . 10^{17}$	0,2 - 1	five beam out reactor
second periphery	22	600	$\leqslant 70$	$0,2 - 3 . 10^{17}$		0,1 - 0,45	two tangential
third periphery	10	600	$\leqslant 70$	$0,6 - 3. 10^{16}$		0,1 - 0,3	one radial axial
fourth periphery	10	600	$\leqslant 70$	$1,5 - 5 . 10^{15}$			two radial lateral
stool (3)	45	600	$\leqslant 70$	$10^{11} - 10^{15}$			

14.3.2	*Reactivity available for experiments:*	*not applicable*
14.3.3	*Neutron spectra (data):*	*See chapter 14.4*
14.3.4	*Specialized equipment available for experimenters:*	*See chapter 14.5*
14.3.5	*Availability of facilities for irradiation on behalf of other organizations of the European Communities:*	*Facilities are available*
14.3.6	*Availability of hot cell facilities:*	*Hot cells are available on the reactor site (see chapter 14.6).*

Fig. 14.1 The TRITON reactor pool.

14.4 INFORMATION SHEET No. 4. NEUTRON SPECTRA (DATA)

The histogram of flux per unit of lethargy for the TRITON reactor is the same as that supplied for the SILOE and MELUSINE reactors, which are tankless pool reactors (see chapters 11.4 and 12.4).

Fig. 14.2 *General view of the working ground of the reactors "TRITON" and "NEREIDE".*

14.5 INFORMATION SHEET No. 5. MAIN CHARACTERISTICS OF SPECIALIZED IRRADIATION DEVICES

14.5.1 Irradiation devices or systems

- *"FRUIT" device for continuous creep studies.*

  This device was designed for studies of the creep in reactors of uranium samples subjected to tensile stress (ref. 1). Its range of use derives from its intended purpose, i.e.

  - *the medium in which the testpiece is submerged consists of a mixture of sodium and potassium;*
  - *the temperature may be set at a predetermined value between 300 °C and 600 °C and can be maintained during reactor shutdowns;*
  - *a tensile force of up to 10 kg can be applied.*

  The deformation rates which one can reasonably expect to study via testpieces having a measuring base about 50 mm long are not less than 10^{-6}/h. The total deformation which can be studied is 20 mm.
  The deformation is measured in accordance with an original procedure for which a patent application has been made (ref. 2). Under this procedure the core of a transducer having two windings (the coefficients of self induction of which vary inversely in accordance with the position of the core) which is rigidly attached to the movable head of the testpiece by a short rod made of molybdenum, a metal having a low coefficient of thermal expansion.
  The displacement of the movable head in relation to the fixed head is measured by a micrometer gauge, of which the essential components, which are submerged in the reactor water, are made of Invar. A 4mm micrometer screw is actuated by a 20 stepping motor capable of operating at 1/4356 revolutions.
  At any given point in time the position of the movable head can be determined from the position of the motor, which corresponds to a given ratio between the values for the coefficients of self induction of the transducer windings connected to the micrometer screw via the Invar chamber. The extension of the testpiece is then measured by converting variations in the differences in length between the testpiece core support rod assembly and the components of the micrometer into the angle of rotation of the micrometer screw.

  It can be seen from the above description that in this device the dimensions of the micrometer components are little affected by variations in heating due to nuclear radiation, so that there is little drift.

- *Systems for irradiation at low temperatures*

  With the appearance on the international market of cryogenerators adapted for the recondensation of cryogenic fluid vapours, it has become possible to modify considerably the design of irradiation systems in swimming-pool reactors. Because of their small size they are easy to install, while their flexibility and the fact that they operate in closed nitrogen, hydrogen or neon circuits ensures a high degree of safety, excellent reliability and economic operation in irradiation experiments.
  The Fontenay-aux-Roses systems (Vinka Loops), which operate at 20 K, are liquid hydrogen cryostats with nitrogen heat insulation, mounted - with their cryogenerators - on a carriage by means of which the bottom of the loop can be moved close to the core, thereby enabling investigations at fluxes of up to 2,2 x 10^{12} n $cm^{-2}s^{-1}$ (fast neutrons).
  *By pressure regulation it is possible to stabilize the temperature to within ± 0.05 K over a temperature range from 20 to 300 K. With these systems, which operate automatically, a total irradiation experience of over 15 000 hours *) has been accumulated, with continuous cycles of as long as 2500 hours.*
  Many different types of experiments can be performed (insitu measurements of length, resistivity, traction, etc.), with the possibility of post-irradiation annealing and removal of samples while cold. All the measurements are done remotely through a data acquisition circuit.

*) *start-up of the first loop (20 K) in 1966*

- *Industrial Neutron Radiography*

This very-high performance installation , which is mounted on a cluster removed from the TRITON reactor (axial channel) is partly automated and used for 90 °/o to satisfy French and foreign industrial needs.

The principal characteristics of this monitoring unit are as follows :

. Collimation ratio	1/200
. Useful range, formats used	18 x 24 cm^2 24 x 30 cm^2
. Total flux density	10^{17} $n.cm^{-2}s^{-1}$
. Thermal flux density ($E \leqslant 0,5$ eV)	6,3 . 10^6 $n.cm^{-2}s^{-1}$
. Gamma dose rate : γA	17 - 27 rad/h
. Max. dimensions of testpieces	0,5 x 0,5 x 0,5 m^3
. Useful length	$\leqslant 4$ m
. Weight of testpieces	$\leqslant 100$ kg
. Exposure range	1' - 20'
. Films currently used	from Industrex A to Industrex B single coated or equivalent

14.5.2 Relevant Publications

- *"FRUIT" device :*

1. C. Thomas de Montpreville
Dispositif "FRUIT" pour l.étude en continu du fluage de l'uranium.
Note CEA N. 1426

2. Demande de brevet No. EN 70 33976.

- *Systems of irradiations at low temperatures :*

1. Les irradiations neutroniques à basses températures dans Mélusine et Siloé.
Irradiations à 20 K dans le réacteur Triton.
Bull. Infs. Scient. Techn. CEA no. 153-154 (Nov.- Dec. 1970), 60 et 66.

2. Depierre, Y.; Verdier, J.
Dispositif d'irradiation à basse température.
Brevet français P.V. 8525 du 9 avril 1965.

3. Conte, R.R.
Irradiations aux basses températures.
Rapport CEA-R-3910 (1969).

4. Conte, R.R.
Dispositif d'irradiation en pile à 20 K,
Cryogenic Engineering Conf., Comm. I, Annexe 1966-5, Boulder (Institut International du Froid, Paris).

5. Conte, R.R.
20 K in pile irradiation facility.
Adv. Cryogen. Engng 12 (1967), 673.

6. Conte, R.R.
Dispositif d'irradiation aux neutrons à 20 K.
Rev. Phys. Appl. 2 (1967).

7. *Weinberg, C.; Conte, R.R.; Dural, J.*
   Dispositif de mesures simultanées de longueur et de résistivité sous irradiation à basses températures.
   Rapport CEA-R-4315 (1972). Note CEA-N-1315 (1970).

8. *Weinberg, C.*
   Croissance de l'uranium sous irradiation à basses températures.
   Rapport CEA-R-4315 (1972).

9. *Conte, R.R.; von Stebut, J.*
   Micromachine de traction fonctionnant aux températures.
   Rev. Phys. Appl. 4 (1969), 70.

10. *Von Stebut, J.*
    Elastoresistivity, A property especially sensitive to stage I radiation defects in copper.
    Phys. Status Solidi (a), 9 (1972) K. 145.

11. *Ponsoye, J.*
    Irradiation de tungstène sous contrainte uniaxiale à basse température.
    Radiat. Eff. 8 (1971), 13.

12. *Lesueur, D.*
    Amorphysation of Pd-Si alloy by irradiation with fission products.
    Fizika (Italy) 2, suppl. 2 (1970), 13.

13. *Leteurtre, J.; Pouchon, J.L.; Zuppiroli, L.; Francois, A.; Housseau, N.; Dural, J.*
    Croissance induite dans le tungstène irradié sous contrainte à 20 K.
    Phys. Lett. 38A 6 (1972), 411.

- *Industrial Neutron Radiography*

1. *Laporte, A.*
   La Neutrographie Industrielle auprès du réacteur TRITON.
   Colloque sur les Applications Modernes de la Pyrotechnic, Toulon, 19 - 22 June 1973.

2. *Laporte, A.*
   Industrial Applications of Neutrons Radiography in France
   Conference on Radiography with Neutrons, Birmingham, 10 Sept. 1973.

3. *Laporte, A.; Mars, J.*
   Two Examples of Quantitative Neutron Radiography
   Conference on Radiography with Neutrons, Birmingham, 10 Sept. 1973.

4. *Bouloumie, J.F.; Laporte, A.*
   Neutrographie Qualitative et Neutrographie Quantitative
   3ème Colloque International sur les Méthodes de Contrôles non Destructifs., Toulouse, May 1974.

5. *Farny, G.; Laporte, A.*
   Industrial Neutron Radiography for Nuclear Uses.
   Euratom Working Group on Irradiation Devices, Harwell meeting, 2 and 3 July, 1974.

14.6 INFORMATION SHEET No. 6. MAIN CHARACTERISTICS OF HOT CELL FACILITIES

Table 14.2: Main characteristics of hot cell facilities

name of laboratory	year of active oper.	scope: active area (m^2)	scope: number of working positions	gamma shielding (material, thickness, density)	leak tightn.	special atmosph.	manipulators number, type (incl. spare manipulators)
TRITON Cell	1962	12	2	heavy concrete 1,03 m, 3,5 g/cm^3	--	--	4 manipulators light-duty CRL-E remote-control handling devices 1 heavy-duty ERTN E 600 handling device
Hot Cell facilities FAR	no data available						

14.6.1 Owner/Operator — TRITON Cell : CEA-CEN, Fontenay-aux-Roses section d'exploitation TRITON

14.6.2 Site — CEN Fontenay-aux-Roses

14.6.3 Associated reactor — TRITON

14.7 *INFORMATION SHEET No. 7. EQUIPMENT AND TECHNIQUES AVAILABLE FOR POST-IRRADIATION EXAMINATIONS.*

- *TRITON Cell*

  This cell is used solely for stripping of material irradiated in the TRITON reactor, i.e., the recovery of testpieces and integrates for shipping to the specialized CEN laboratories at Saclay, and the disposal of radioactive waste.

  The cell equipment comprises :
  - *1 stripping millet*
  - *1 cutting machine for use under a controlled atmosphere*
  - *1 ac electric saw*
  - *1 set of hydraulic shears*
  - *1 pneumatic vice, plus pneumatic tools.*

- *Hot Cell Facilities FAR :* *no data available*

- *Relevant publications :*

  - *PILE TRITON*
    Rapport descriptif, Sept. 1967, CEA No. 840.
  - *Demantèlement "COLIBRI T" en tronçonneuse étanche pour mesure de pression et analyse des produits de corrosion.*
    Michel Faure
    DPE/SET No. 169/70.

14.8 *INFORMATION SHEET No. 8. UTILIZATION AND SPECIALIZATION OF THE HOT CELL FACILITIES*

- *TRITON Cell :* *see chapter 14.7*
- *Hot Cell Facilities FAR :* *large hot laboratory complex, specializing basically in the examination of Pu pins (RM2 system).*
- *Availability of hot cell facilities for post irradiation examinations on behalf of other organizations within the European Communities.*

  TRITON Cell is available; other hot cell facilities are not available.

15. THE ESSOR REACTOR AT ISPRA, ITALY AND AVAILABLE HOT CELL FACILITIES. INFORMATION SHEETS.

Fig. 15.1 The ESSOR reactor on the site of the Ispra Establishment of the Joint Research Centre.

15.1 INFORMATION SHEET No. 1. MAIN CHARACTERISTICS OF THE REACTOR.

Name of reactor	*ESSOR*
Organization	*Commission of the European Communities, Joint Research Centre, Ispra Establishment*
Site	*Ispra (Varese), JRC*
In operation since	*December 1968*
Power MW (th)	- *25 in driver zone* - *up to 18 MW in experimental zone*
Moderator	D_2O *at 99,74* $^o/o$
Coolant	- D_2O *in driver zone* - H_2O, D_2O *in experimental zone*
Fluid and pressure in experiment channels	*see chapter 15.2*
Nominal operation (days/year)	*~ 200*

Operating cycle (days)	*typical*	*long*	*short*
on power		*50*	*22*
shut down		*8*	*4*

Main features of the reactor are summarized in Table 15.1

Responsible officer (name, address):

ESSOR and Hot Cells: *T. Doyle* *Commissione delle Comunità Europee*
Centro Comune di Ricerca
Stabilimento di Ispra, Divisione ESSOR
I - 21020 Ispra (Varese)

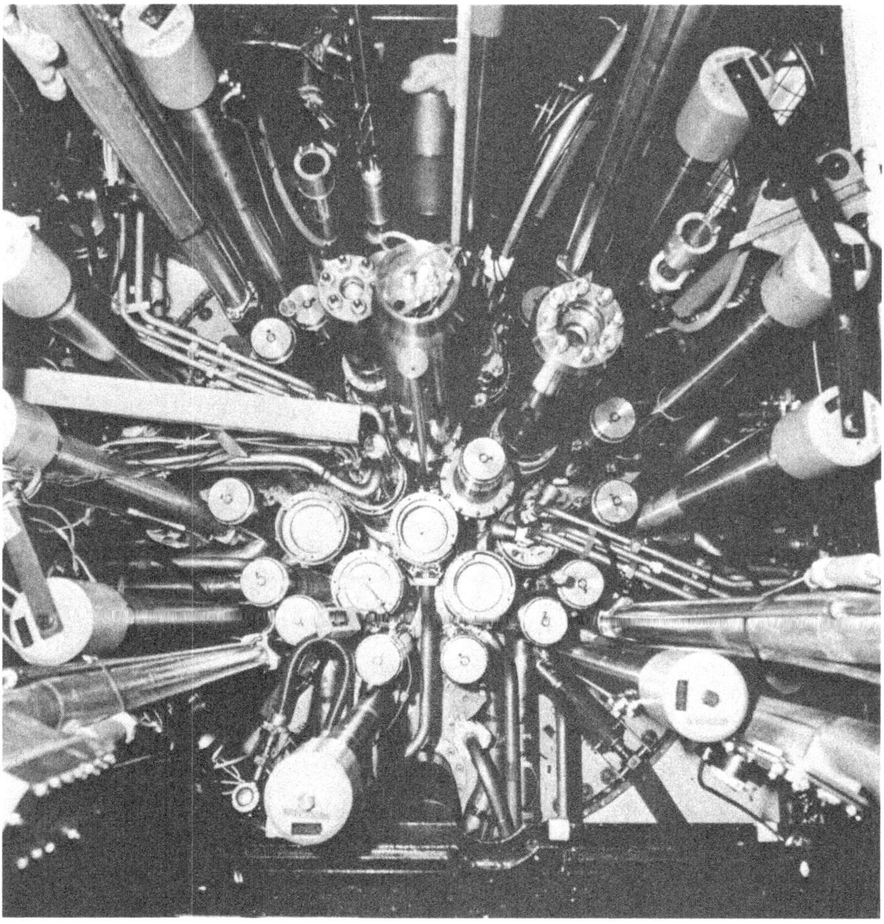

Fig. 15.2 *ESSOR Upper Reactor Room.*

Table 15.1 : Main features of the reactor

Core	*Active dimensions*	*diameter 118 cm* *height 154 cm*
	Sub-assemblies	*16 feeder fuel elements of 6 concentric plates containing 465 or 650 g U-235 each.*
		12 experimental zone positions, maximum diameter 17 cm.
		4 peripheral zone positions, maximum diameter 7,5 cm.
	Moderator	*12.000 kg D_2O at 50 ^oC and atmospheric pressure.*
Control	*Safety rods*	*4 mechanical type, B_4C* *(6000 pcm negative reactivity in 1 s)*
		6 liquid poison type, Li_3BO_3 *(7000 pcm negative reactivity in 0,35 s)*
	Shim rods	*8 mechanical type, Cd* *(12 000 pcm negative reactivity at 7 pcm/s)*
	Control rods	*2 mechanical type, stainless steel* *(8 00 pcm negative reactivity at 40 pcm/s)*
Pressure Vessel	*Material*	*AISI 304 L stainless steel*
	Dimensions	*diameter 238 cm* *height 423 cm*
	Design pressure	*8 kg/cm^2*
Handling	*Feeder zone machine*	*Maximum capacity*
		diameter 14 cm *height 573 cm*
	Experimental zone machine	*Maximum capacity*
		diameter 29 cm *height 935 cm*

15.2 INFORMATION SHEET No. 2 UTILIZATION AND SPECIALIZATION OF THE REACTOR

15.2.1 Original purposes of the reactor

- *Technological study of Orgel type reactor : study of organic coolant; pressure and calandary tube, S.A.P. cladding material, U.C. fuel elements*
- *Development of CIRENE type fuel*

15.2.2 Major modifications subsequently made

- *Some organic loops were dismantled and substituted by LW loops*
- *Some specific devices have been added (other indications in successive sheets)*
- *Work performed in the framework of a contract between the Commission and the Italian Government.*

15.2.3 *Major modifications planned*

None

15.2.4 *Major projects and programmes for which the reactor presently is used*

- *Testing of fuel elements for CIRENE type reactor*
- *Testing of liquid safety rods.*
- *Special irradiation devices.*
- *Construction of Super Sara circuit for in-pile fuel behaviour studies under LOCA conditions.*

15.2.5 *Loops and irradiation devices specially developed for the reactor*

- *CART TC loop :*	*light water fog cooled loop for testing CIRENE fuel assemblies.*
- *SARA loop :*	*(2.26 MW, 180 ata), for irradiation of LW reactor fuel pins in steady state and transient conditions (mainly LOCA) for safety assessment.*
- *Zenon devices:*	*irradiation at low temperature in the inner cavity of the driver fuel elements.*
- *MODESTE devices:*	*simultaneous irradiation of four capsules or instrumented fuel-pins in the experimental zone.*
- *RABBIT:*	*irradiation of non-instrumented capsules or fuel pellets in the experimental zone with the possibility to remove the experiment at any time.*
- *Liquid safety rods:*	*testing under irradiation of various components of the system.*
- *SIDRONE:*	*automated irradiation facility for neutron transmutation doping of large silicon ingots (Φ 76,2; l. 500mm) for semiconductors.*
- *EOLO Jr :*	*instrumented fuel rod for in-pile clad ballooning studies.*
- *COMMODE :*	*irradiation of samples of pressure and calandria-tubes for CIRENE reactor.*

15.3 *INFORMATION SHEET No. 3. EXPERIMENTAL FACILITIES OF THE REACTOR*

15.3.1 *Irradiation positions*

Table 15.2: Irradiation positions

type/name	*number*	*active length mm*	*useful diam. mm*	*typical unperturbed neutron flux density x 10^{18} m^{-2} s^{-1}* thermal	fast (E > 0,1 MeV)	*nuclear heating W/g Al*
experimental channel	*12*	*2200*	*170 *)*	*3 - 4*	*0,05*	*~ 0,5*
driver zone channel	*16*	*1540*	*30,7 **)*	*1,5*	*1,1*	*~ 5*
reflector zone channel	*4*	*2200*	*78 *)*	*1,5*	*0,01*	*~ 1*

**) facility hole diameter*
***) driver fuel element central hole (for Zenon)*

15.3.2	*Reactivity available for experiments:*	*4000 pcm*
15.3.3	*Neutron spectra (data):*	*see chapter 15.4*
15.3.4	*Specialized equipment available for experimenters:*	*clean workshop, analytical chemistry laboratory, Hot Laboratories for dismantling and post-irradiation examination of experiments.*
15.3.5	*Availability of facilities for irradiation on behalf of other organizations of the European Communities:*	*Facilities are available*
15.3.6	*Availability of hot cell facilities:*	*Hot cell facilities are available on the reactor site:* *ADECO, ATFI and LMA* *For details see chapter 15.6*
15.3.7	*Relevant publications:*	- *The ESSOR Plant Facility (Sept. 1978)* - *The Super-Sara Programme in ESSOR.* - *SIDRONE: An automated Irradiation Facility for Neutron Transmutation Doping of Silicon-Crystals.*

15.4 INFORMATION SHEET No. 4. NEUTRON SPECTRA (DATA)

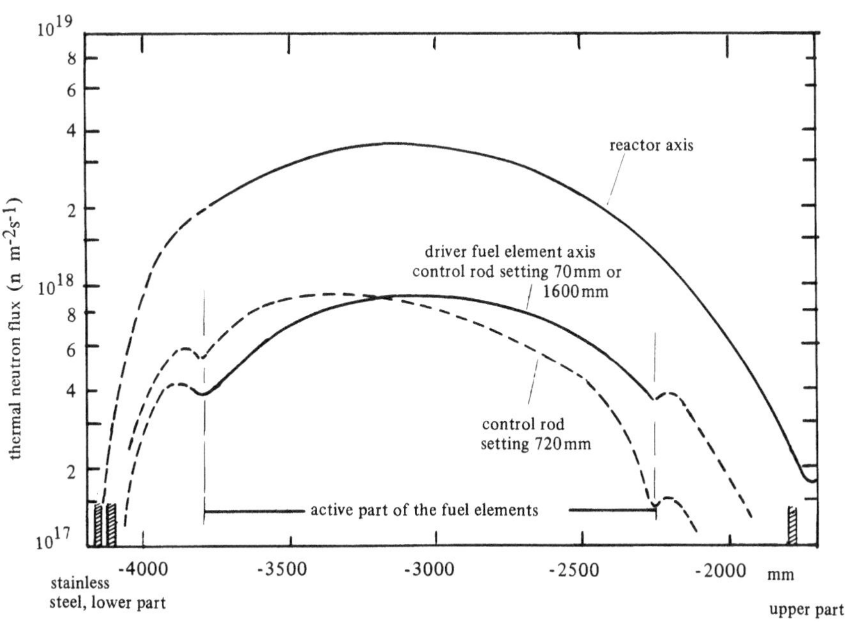

Fig. 15.3 Neutron flux distribution across the ESSOR reactor core. Axial distribution.

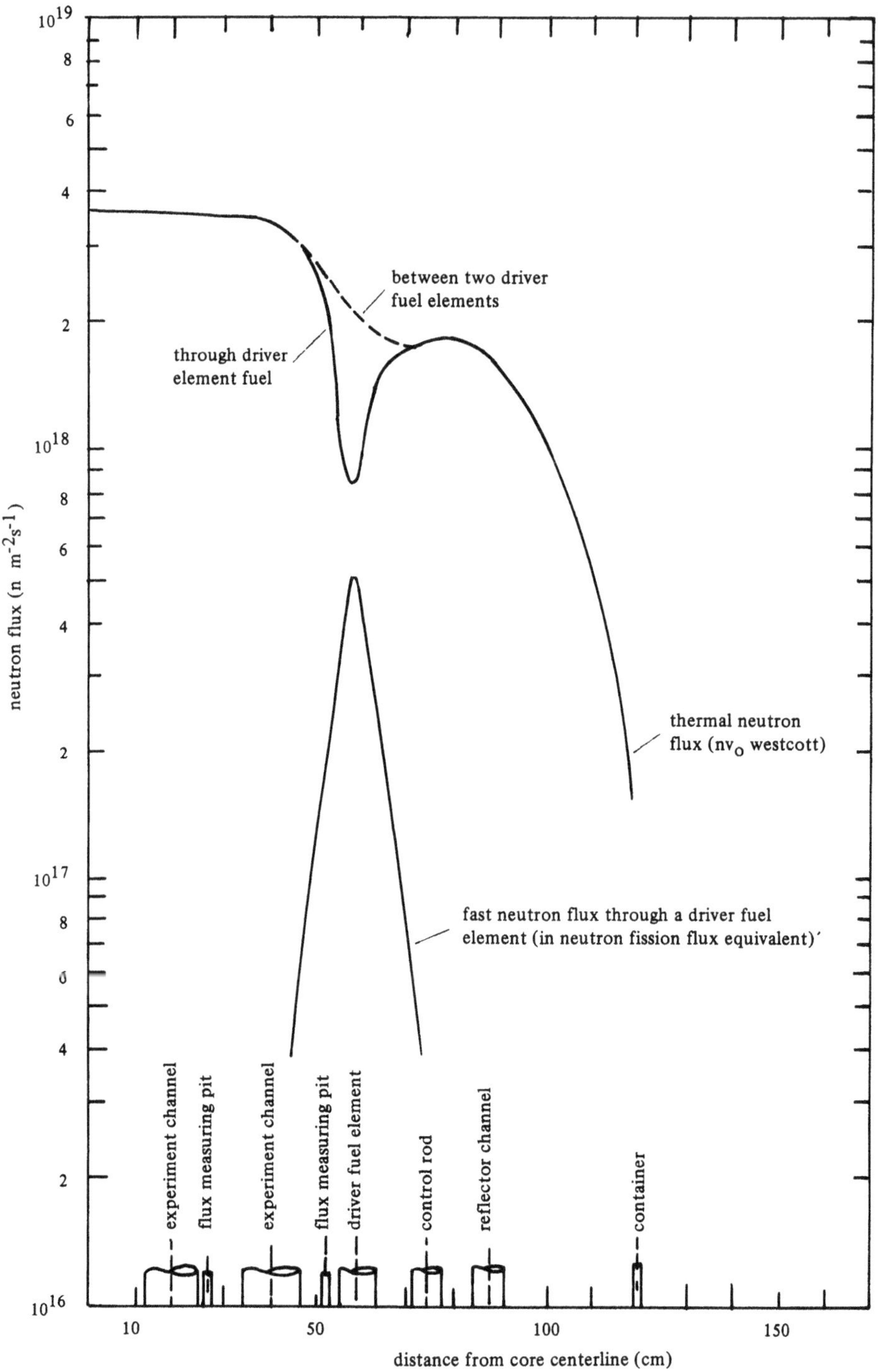

Fig. 15.4 Neutron flux distribution across the ESSOR reactor core. Radial distribution at mid-level.

15.5 INFORMATION SHEET No. 5. MAIN CHARACTERISTICS OF SPECIALIZED IRRADIATION DEVICES AND LOOPS.

Table 15.3: Main characteristics of experimental devices and loops.

devices	dimensions		unperturbed neutron flux in $10^{18} m^{-2} s^{-1}$		axial form factor	max. power MW	coolant					
	diameter cm	length cm	thermal	fast ($E >$ 0,1 MeV)			nature	pressure max. kg/cm^2	loss of pressure kg/cm^2	flow m^3/h	inlet temp. ^oC	outlet temp. max. ^oC
Devices												
Zenon (Cirene)	3,07	160	1,5	1,1	0,75	0,1	D_2O	8	3	89	45	60
MODESTE 4	17	220	3 - 4	0,05	0,65	0,25	D_2O	1,2	1,2	16	45	80
MODESTE Gioconda	5,4	1,50	3,5	0,05	0,80	0,06	D_2O	1,2	1,2	6	45	47
Rabbit Radioelement	2,6	10	3	0,05	0,9	–	D_2O	~ 3	--	5	45	80
Commode	16	120	0,8	0,33	0,85	2,5	D_2O	8	--	90	45	80
SIDRONE	7,65	50	1,7	0,05	0,96	--	D_2O	1,2	--	5	45	60
Loops												
CART TC	10,6	220	3	0,05	0,65	1,5	fog	50	1,5	10 *)	250	264
SARA	11,3	200	3	0,05	0,7	2.26	H_2O	60	--	12,0 *)	300	330

*) $kg s^{-1}$

15.6 INFORMATION SHEET No. 6. MAIN CHARACTERISTICS OF HOT CELL FACILITIES

Table 15.4: Main characteristics of hot cell facilities

name of laboratory	year of operation	scope: active area/ active volume	scope: number of cells/ number of working positions	gamma shielding (material, thickness and density)	leak tightn.	special atmosph.	manipulators number and type	main characteristics
ADECO	1970	110 m^2 / 520 m^3	6 / 9	heavy concrete, 1,1 m; 3,5 g/cm^3	yes	air	18 M.S. extended reach Central Research L, type E and F 1 Electromechanical telemanipulator weight : 3 kN all positions, 10 kN lifting	direct connection with ESSOR reactor Connections with LMA and ATFI
ATFI	–	80 m^2 / 580 m^3	3 / 4	concrete, 1.1 m, 2,3 g/cm^3	no	air	6 M.S. extended reach Central Research L type E and F 1 Electromechanical telemanipulator weight 1 kN	only for non-fissile materials Direct connection with ESSOR reactor.
LMA	1968	~ 40 m^2 / ~ 130 m^3	20 / 22	heavy concrete, 1 m; 3,4 g/cm^3 lead, 15 cm	yes	air	2 M.S. extended reach Central Research L Mod. E HD	Electronic microscope
							4 M.S. Nuclear Equipment L Mod. M 9	metallographic active Lab.
							4 MA 11 CEA SGDG / La Calhène 18 M7 OBSON La Calhène	mechanical test cells

15.6.1 Owner /Operator — *Commission of the European Communities, Joint Research Centre, Ispra Establishment*

15.6.2 Site — *Ispra (Varese), JRC*

15.6.3 Associated reactor — *ESSOR*

15.7 INFORMATION SHEET No. 7. EQUIPMENT AND TECHNIQUES AVAILABLE FOR POST-IRRADIATION EXAMINATIONS

15.7.1 ADECO Facility

- *under water dismantling device for PWR-type fuel assembly*
- *dismantling rig equipped with longitudinal, transversal cutting, extremity milling*
- *8 dry storage and 1 washing pits*
- *visual and photographic examination, presentation bench*
- *shuffling device for CIRENE fuel bundles*
- *fuel pin cutting device*
- *neutron radiography examination device (not in service)*
- *gamma spectrometry vertical bench with GeLi detector and computer programme*
- *cladding failure detection device, bubble test system in oil at 270 °C.*
- *bench-scale vitrification installation, ESTER programme*

15.7.2 Relevant publications

- *J. Biteau, C. Garric, A. Wilhelen*
  Dismantling workshop for Orgel fuel elements
  IAEA International Symposium on Working Methods in High Activity Hot Laboratories, June 1965 - Paper no. 44.
- *Facilities at the JRC Ispra for Manipulation and Post-Irradiation Examination of Fuel Elements for Power Reactors.*
  15th Meeting of Euratom Hot Laboratories Committee, Geesthacht, June 1973.
- *Post-Irradiation Analysis of TRINO Vercelles Reactor Fuel Elements.*
  EUR 4909e, 1972.

15.8 INFORMATION SHEET No. 8. UTILIZATION AND SPECIALIZATION OF THE HOT CELL FACILITIES (ADECO)

15.8.1 Original purposes of the facility

Dismantling and non destructive examination of Orgel fuel elements

15.8.2 Major modifications subsequently made

The devices were adapted for Cirene fuel elements type and for dismantling of specific rigs.
Improvement of the devices for LWR assemblies examination.

15.8.3 Major modifications planned

- *Equipment of one cell with other non destructive testing device as Eddy current test and macrophotographic devices for cladding examination.*
- *Equipment for destructive examination of heavily damaged fuel assemblies for Reactor Safety Program (S.S.T.P.)*
- *Equipment for liquid waste disposal.*

15.8.4 Major projects and programmes for which the facility presently is used

- *Cirene fuel elements post-irradiation examination*
- *Support to the Power LW reactor technology*
- *Dry storage test*

15.8.5 Availability of hot cell facilities for post irradiation examinations on behalf of other organizations within the European Communities

- *Facilities are available.*

Fig.15.5 View of the ADECO Hot Cells (front zone).

16. THE HIGH FLUX REACTOR (HFR) AT PETTEN, THE NETHERLANDS AND ITS ASSOCIATED HOT CELL FACILITIES. INFORMATION SHEETS.

Fig. 16.1 HFR Petten. Reactor and annexed buildings

16.1 INFORMATION SHEET No. 1. MAIN CHARACTERISTICS OF THE REACTOR

Name of reactor	*HFR - High Flux Reactor*
Organization	*Commission of the European Communities, Joint Research Centre, Petten Establishment*
Site	*Petten, JRC*
In operation since	*1961*
Power (th)	*45 MW *)*
Moderator	H_2O, *Be*
Coolant	H_2O
Fluid and pressure in experiment channels	H_2O, *3,6 bar abs.*
Nominal operation (days/year)	*ca. 280*
Operating cycle (days)	
on power	*25.7*
shut down	*2.3*

**) the reactor scram level is set at 50 MW*

Main features of the reactor :

- *reactor core* — *The core lattice is a 9 x 9 array (729mm x 693mm) containing 33 fuel assemblies, 6 control members, 17 experiment positions and 25 beryllium or aluminium reflector elements. Height of the core 924 mm.*

- *fuel assemblies* — *The fuel elements (horizontal cross section 81 mm x 77mm, height 924mm) contain 23 vertically arranged curved fuel plates with a height of 625 mm. The length of the fuel inside the plate is 600mm. The uranium is about 93 $^{o}/o$ enriched in ^{235}U. Each fresh fuel assembly contains 405g ^{235}U and 1000mg ^{10}B. The boron is concentrated in the two flat side plates of the fuel assemblies.*

- *control members* — *The control members consist of a cadmium section on top of a fuel section, which contains 19 fuel plates with a total fresh mass of 280g ^{235}U. Control members are moved upwards during a cycle but only over a small range, i.e. variation with time of vertical distribution of neutron flux density and nuclear heating during an operation cycle is small.*

- *reactor vessel* — *The reactor vessel, which contains in its lower part the reactor core, is a welded 5154 aluminium tank, 1.60m in diameter, 5.4m high and 25mm thick. Internal overpressure is 260 kPa. The cylindrical reactor wall is interrupted at one side, providing a rectangular compartment, the height and width of which are identical to those of the core size. This pool side facility is accessible from the reactor pool for installing irradiation devices. The distance of an irradiation device from core box wall is variable, so that the pool side facility is especially suited for fuel irradiations (e.g. power ramp experiment).*

- *reactor cooling* — *Reactor cooling is provided by recirculating demineralized water in a primary closed circuit. Flow rate through the core is 4100 m^3/h. Water inlet temperature is about 313 K, corresponding outlet temperature is 323 K. Reactor power is dissipated in a water-to-water heat-exchanger to the open secondary cooling system.*

Responsible officers (name, address):

- *HFR:* — *P. von der Hardt*
  Joint Research Centre, Petten Establishment
  P.O. Box 2
  1755 ZG Petten, the Netherlands

- *Hot Cells - LSO :* — *H.J. Wervers*
  Netherlands Energy Research Foundation - ECN
  P.O. Box 1
  1755 ZG Petten, The Netherlands

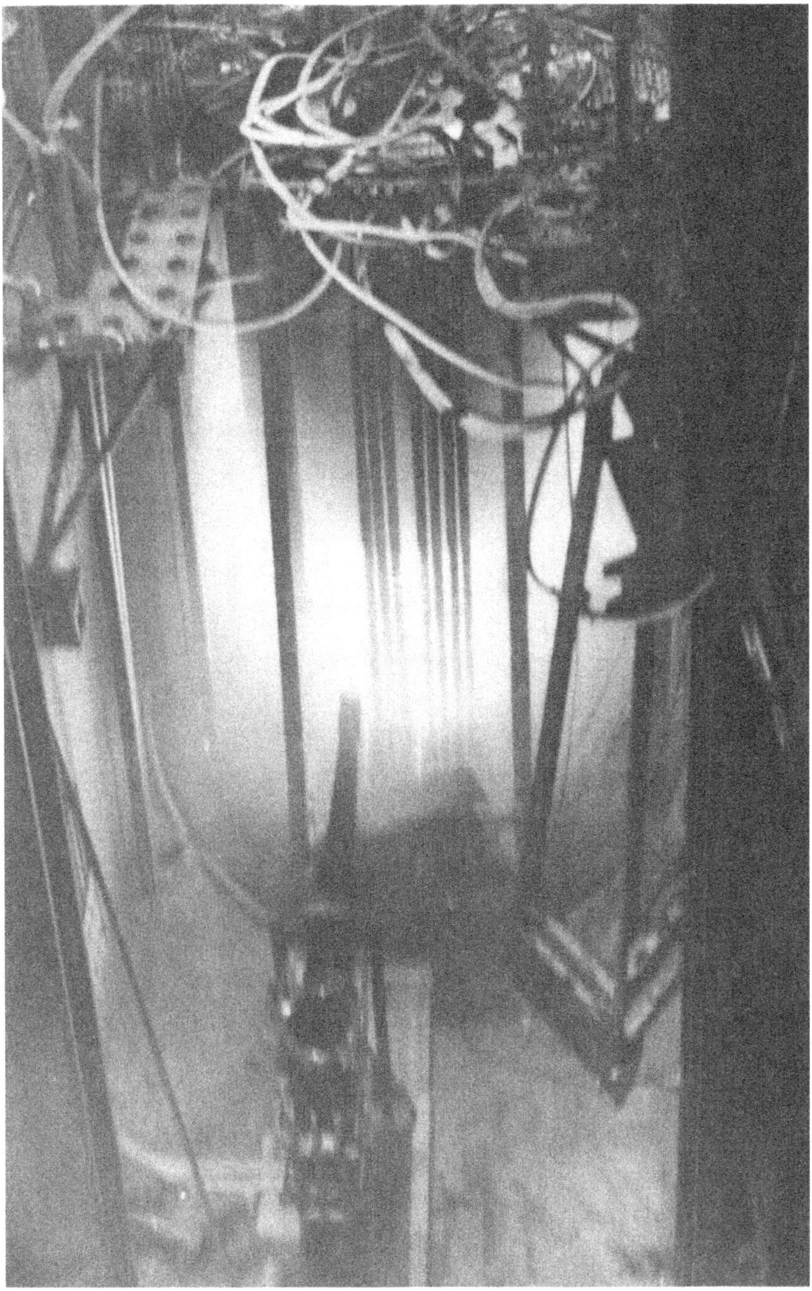

Fig. 16.2 HFR Petten. Total view of the pool side facility.

16.2 INFORMATION SHEET No. 2 UTILIZATION AND SPECIALIZATION OF THE REACTOR

16.2.1 Original purposes of the reactor

- *Materials testing*
- *Isotope production*
- *Beamhole experiments (solid state and nuclear physics)*

16.2.2 Major modifications subsequently made

- *Power increases 20 to 30 and 30 to 45 MW*
- *Introduction of burnable poison fuel*
- *Several core configuration changes*
- *New in-tank experiment penetrations*
- *New PSF support table with increased spacing between neighbouring irradiation positions*
- *Complete replacement of reactor- and general purpose experimental instrumentation*
- *Several improvements on major plant systems*
- *Extension of annexed buildings.*

16.2.3 Major modifications planned

- *Renewal of reactor instrumentation in 1981 / 83*
- *Replacement of reactor vessel in 1983*
- *Power increase to 60 MW (after 1983 , studies)*
- *New multi-purpose workshop and storage building (1984 / 85)*

16.2.4 Major projects and programmes for which the reactor presently is used

- *Fast breeder reactor structural materials irradiations*
- *Fast breeder reactor fuel under operational transients*
- *Light water reactor fuel pin power ramping*
- *High temperature gas cooled reactor graphite and fuel element irradiations*
- *Nuclear structura and solid state physics experiments*
- *Radioisotope production, activation analysis*
- *Neutron radiography, neutron dosimetry development*

16.2.5 Loops and irradiation devices specially developed for the reactor

Various capsules for the above-mentioned programmes (see chapter 16.5 for names and main characteristics)

16.3 INFORMATION SHEET No. 3. EXPERIMENTAL FACILITIES OF THE REACTOR

16.3.1 Irradiation positions

Table 16.1: Irradiation positions

type/name	number	active length mm	useful diam. mm	maximum neutron flux density ($m^{-2}s^{-1}$) thermal	fast ($E > 0.1\,MeV$)	max. nuclear heating W/g Al
in-core positions	9	600	70	$2 \cdot 10^{18}$	$4{,}5 \cdot 10^{18}$	11,9
reflector positions	8	600	70	$1{,}8 \cdot 10^{18}$	$2{,}5 \cdot 10^{18}$	5,7
pool side facility (PSF)	10	500	50	$3{,}3 \cdot 10^{18}$	$0{,}8 \cdot 10^{18}$	3,3
beamholes	11	-	175 (8) 250 (2) 630 (1)	$10^9 - 10^{13}$	--	--

16.3.2 Reactivity available for experiments: $2 - 3\ ^0/o\ \Delta k/k$ *)

16.3.3 Neutron spectra (data): see chapter 16.4

16.3.4 Specialized equipment available for experimenters:

- neutron radiography
- microprocessor-based experiment data acquisition system
- centralized data processing system (DACOS)
- out-pile gas installation for simultaneous operation of 8 double-wall fuel capsules
- out-pile installation for control and operation of fuel pin ramping experiments
- sweep gas loops for operation of HTR fuel capsules
- various gas panels for operation and temperature regulation of REFA and TRIO reloadable facilities

16.3.5 Availability of experimental facilities :

The facilities are available for irradiations on behalf of other organizations of the European Communities and also for third parties.

16.3.6 Availability of hot cell facilities:

Hot cell facilities are available on the reactor site (see chapter 16.6)

16.3.7 Relevant recent publications:

- Annual report 1979, EUR 6818 EN FS
- HFR Petten. Characteristics of Facilities and Standard Irradiation Devices, EUR 5700e (edition 1981)

*) available for the 17 in-core experiment in relation to the situation where all experiment positions (17) are loaded with Al- filler elements.

16.4 INFORMATION SHEET No. 4. NEUTRON SPECTRA (DATA)

The neutron spectrum has been determined in the core positions C3, E5, F2, G5, G7, H4 and H8 of the standard core, and in six positions of the pool side facility .
The activated data from foils irradiated in these positions were used as input in the computer programme SAND-II. Using this programme the neutron spectra for the positions have been calculated for the energy range 10^{-10} to 20 MeV. Several neutron spectra at 45 MW are given in Figs. 16.3 - 16.7.

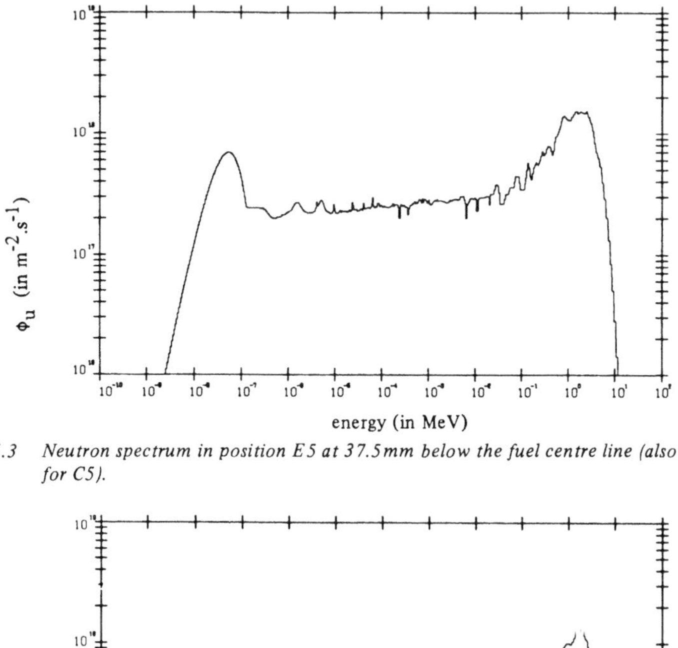

Fig. 16.3 Neutron spectrum in position E5 at 37.5mm below the fuel centre line (also representative for C5).

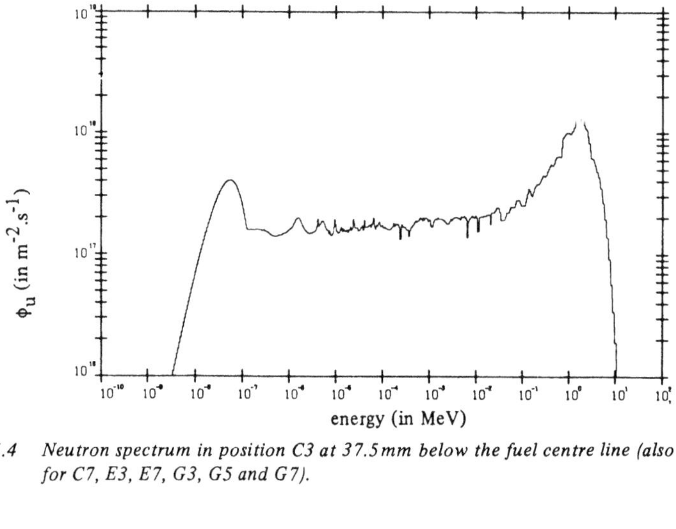

Fig. 16.4 Neutron spectrum in position C3 at 37.5mm below the fuel centre line (also representative for C7, E3, E7, G3, G5 and G7).

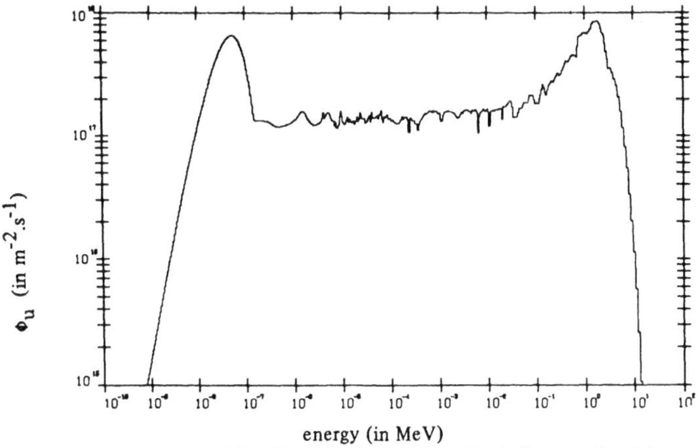

Fig. 16.5 *Neutron spectrum in position F2 at 37.5 mm below the fuel centre line (also representative for D2, D8, F8, H4 and H6).*

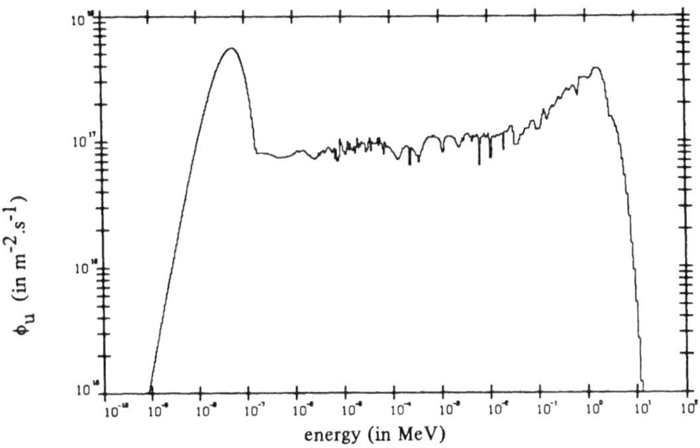

Fig. 16.6 *Neutron spectrum in position H8 at 37.5 mm below the fuel centre line (also representative for H2).*

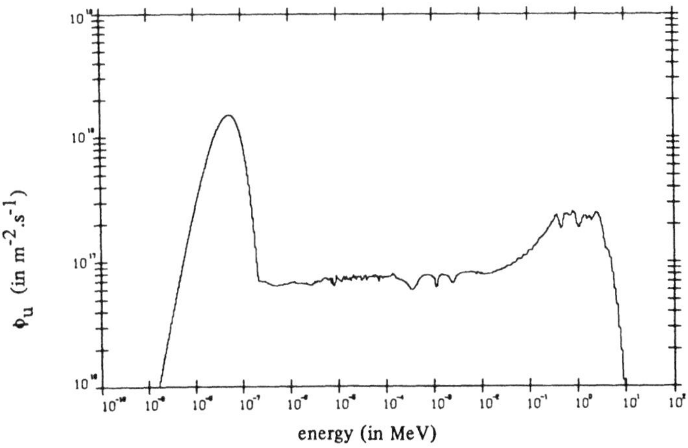

Fig. 16.7 *Neutron spectrum in PSF6, at 25mm from the core box wall and at 37.5mm below the fuel centre line (also representative for the other five PSF positions in which measurements have been done).*

16.5 INFORMATION SHEET No. 5. MAIN CHARACTERISTICS OF SPECIALIZED IRRADIATION DEVICES

16.5.1 Irradiation devices for fissile materials

- *FSC rig*

  Irradiation of full size cylindrical HTR samples up to 200 kWm^{-1} (diameter dependent) at 675K - 1475K sample surface temperature.
  Useful length/useful diameter 600 mm / 63 mm

- *BEST rig*

  Irradiation of spherical HTR fuel elements and/or segments from block fuel elements up to 200 kWm^{-1} at 675 - 1475 K specimen surface temperature.
  Useful length / useful diameter: 600mm/60mm.

- *BWFC*

  Irradiation testing of LWR fuel pins under BWR/PWR conditions. Heat removal from specimen by sub-cooled local boiling in capsule pressure vessel. Variable fuel pin power density (cycles, ramps) by horizontal displacement unit. Measurement of total power output, pin or fuel elongation, centre temperature and fission gas pressure.
  Reloadable under water to test pre-irradiated fuel pins. Linear heat rating: about 100 kWm^{-1};
  Maximum coolant temperature: 625 K;
  Specimen length/diameter: 400mm/10mm.

- *ELLAS (EXOR) type capsule*

  Irradiation of fuel pins in sodium or NaK at heat ratings up to 120 kWm^{-1}. Double-wall capsule with small gas gaps between the walls. Temperature regulation by gas mixture. Variable fuel pin power by horizontal displacement unit.
  Specimen length/diameter: 400mm/5 to 20 mm (max. to 60mm under certain conditions)

  Special ELLAS type capsules for LMFBR fuel pin irradiations are:

  CAREL

  Irradiation (cycling, ramping) of high enriched, mixed UPuC fuel pins at 80 kWm^{-1} linear power and 895 K cladding temperature. Capsule with Cd-screen.
  Fuel pin length/diameter: 442mm / 8,5mm.

  DUELL

  Double-wall capsules for irradiation of two parallelly arranged short FBR fuel pins.
  The DUELL experiments are start-up ramp tests between 40 and 80 kWm^{-1} maximum linear power at mid-clad temperature between 675 K to 875 K.
  Fuel pin length/diameter: 435mm/6mm (7,6mm)

  KAKADU

  Double-wall capsules for irradiation of two parallelly arranged long FBR fuel pins.
  The KAKADU experiments are in-situ ramp tests between 40 and 80 kWm^{-1} maximum linear power at mid-clad temperatures between 675K and 875K.
  Fuel pin length/diameter: 1556 / 6 (7,6) mm.

- *TRIO and REFA reloadable facilities*

  Irradiation of fissile and non-fissile materials under various conditions. Standard reloadable facilities with one (REFA) or three (TRIO) parallel channels. Vertical displacement unit for axial temperature adjustment. Linear heat rating up to about 100 kWm^{-1} (TRIO) or 150 kWm^{-1} (REFA); specimen temperature range: 775 - 1775K.
  Useful length/diameter: 880mm / 40mm or 70mm (REFA) ;
  880 mm / 29mm or 31,5mm (TRIO).

Instrumented inserts (sample holders) are largely standardized.
Special sample holder using TRIO and REFA type facilities are :

ARTEMIS

Sample-holder for REFA-70 to study coated particle failure in a high temperature gradient field. Nominal temperature: 1575K; temperature gradient 575K/cm

BATAVIA

Sample-holder (TRIO) to study irradiation behaviour of a large variety of coated particle fuel to a high burn-up at temperatures of 1275 ± 50K; 1525 ± 50K; 1775 ± 50K (various sample holders).

CATRI

Sample-holder for special TRIO with Cd-screen to irradiate (steady-state overload conditions) high-enriched, mixed UPuC fuel pins at a maximum linear power of 100 kWm^{-1} *and central temperatures of 2500 K. Cladding temperature : 925 K;*
Fuel pin length/diameter: 478mm / 8,5mm.

- *LOC*

  Capsule for loss of cooling accident simulation. Triple contained fuel pin. Loss of cooling initiated by increasing radial heat resistance. Temperature regulation by variation of gas mixtures. Loss of cooling transients after certain burn-ups. Programmed thermal cycling. Linear heat rating: 600 kWm^{-1}*; cladding temperature: 800 K normal and 1325K loss of cooling.*
  Specimen length/diameter: 250 mm / 6mm.

- *SHOT*

  Double-wall capsule for simulation of local blockage in fast neutron assembly. Simulation of total blockage by increasing fuel pin pressure. Pressure difference over the cladding: 2 - 10 MNm^{-2}*. Temperature control by variation of gas mixture. Programmed thermal cycling. Linear heat rating: 60 kWm*$^{-1}$*; specimen surface temperature: up to about 1125 K.*

- *TOP*

  Capsule for mild transient overpower experiments with LMFBR fuel pins. Overpower realization by controlled decrease of BF_3 *pressure in annulus around fuel pin. Overpower factor 1 to 4. Fuel elongation, fuel pellet temperature, fission gas pressure build-up are measured during power transients. Maximum linear power (normal) 42 kWm*$^{-1}$*.*
  Fuel pin length/diameter: 355mm / 7,6mm Enrichment of UO_2 *fuel: 20 °/o U-235.*

- *POCY*

  Pool side facility (chimney) capsule for power cycling of mixed carbide fuel pins between 60 and 100 kWm^{-1} *at clad temperatures between 675K and 775K. Horizontal displacement unit (limited). Fuel pin profile is measured at intervals by lifting the fuel pin and passing it through an incorporated diameter gauge.*
  Fuel pin length/diameter: 550 mm / 9,5mm.

16.5.2 *Irradiation devices for structural materials*

- *TRIO and REFA reloadable facilities (see chapter 16.5.1).*
  Special TRIO or REFA sample holders (inserts) are:

  AUSTIN

  Sodium filled sample holder (TRIO) for long term irradiation (damage rate 30 dpa) of small austenitic steel specimens for strain rate studies. Temperature range 775 K to 825 K.

  CRIMP

  Graphite creep facility (REFA) for continuous tensile creep measurement on a graphite sample. (length/diameter: 152mm/20mm), stressed by pressurized bellows.
  Specimen surrounded by unstressed reference shell. Differential strain measurements by inductive transducer and bellows actuated sensor. Stepped gas gap.
  Temperature control by binary gas mixture. Total gauge length: 100mm;
  Tensile load: 0,6 kN; Temperature range: 875K to 1475K.

  DISCREET

  Sample holder for TRIO-facility to study tensile or compressive creep behaviour of graphite under irradiation. Graphite specimens are removable for intermittent deformation measurements during reactor shut-down periods. Graphite specimens stressed by pressurized bellows are surrounded by unstressed reference sheels. Temperature regulation by binary gas mixture. Tensile load: 0,7 kN; compressive load: 2 kN. Temperature range 475K to 975K. Number of samples per holder: 5 tensile specimens (Ø18mm x 50mm) or 10 compressive specimens (Ø26mm x 40mm).

- *Standard graphite rigs*

  TRIO sample holder for unstressed graphite irradiations in inert atmosphere. Depending on the irradiation temperature only, these are the following standard sample holders.

  type 1 : all aluminium sample holder for temperatures up to 725K.
  type 2 : graphite drums in stainless steel tube, for 775K and 875K.
  type 3 : graphite drums in stainless steel tube for 975K and 1075K.
  type 4 : graphite drums in TZM tube for 1175 K and 1275 K.
  type 5 : all metal carrier made from TZM or Nb for 1375 K and 1475K.

  useful length/diameter: 880mm / 29 or 31,5mm.

  FANTASIA

  Sodium filled sample holder (TRIO) with stainless steel samples to study fracture toughness under LMFBR reactor operation conditions.
  Sample temperature: 625K and 825K; number of samples per holder: 4 tensile (2,5 x 10 x 30mm) and 4 3PB samples (15x20x 110mm).

  NIRVANA

  Sample holder (TRIO) for irradiation of Nb and V (candidate fusion reactor materials) at 725 K, 875 K, 1000 K and 1125 K. Special attention is given to the fact that the sample material is very sensitive to any pollution of its environment in O_2, N_2, H_2 or C_2.

  SINAS

  Sodium filled sample holder (TRIO) for irradiation of 15 stainless steel CT samples (size Ø 10mm x 60mm). Irradiation temperature 825 K ± 20K.
  Up to now 100 sample carriers of this type have been built and were irradiated in the frame of a fracture propagation research programme.

- *HOBBIE*

Creep facility for continuous deformation measurements of tubes (Zircaloy). Tubular test specimens are stressed by external gas pressure. Deformation measurements by up to 20 eddy current coils. Test speciment temperature control by nuclear heating plus electrical heaters Simulation of radial fuel-to-cladding gas gap and pellet-to-pellet gap by beryllium and nickel sleeves respectively. On-line measurement of deformation, external pressure and temperature Pressure differential across the tube wall: up to 20.7 MPa; deformation measurement range: 0 to 1,5 mm; temperature range 575K to 705K; specimen length/diameter: 150mm/11mm.

- *MONA*

The capsule contains 9 steel samples in a single horizontal layer inside a sodium-filled, helium isolated sample holder. It is specially designed for very short irradiations (hours) at exact and constant temperature. The capsule can be loaded and unloaded with the reactor at full power. Equilibrium temperatures are reached within 5 to 10 minutes. Fast and accurate temperature control is achieved by vertical displacement. Typical sample temperature 823 ± 25 K.

- *NAST*

Sodium filled, double-wall irradiation rig to study neutron embrittlement of a large number (42) of tensile stainless steel specimens. Inner and outer heat removal. Accurate temperature in space and time. Temperature range 725K to 825K.
Useful length/diameter: 525mm / I.D. 27mm, O.D. 54mm.

16.5.3 Radionuclide Production Facility

- *RIP*

The Reflector Isotope Plug consists of 8 aluminium trays connected to a central rod. Each tray is designed to hold one 25mm Ø capsule, one 16mm Ø capsule and three 10mm Ø capsules of 80mm length (external dimensions). Presently not in use.

- *RIF*

A simple holder (Reloadable Isotope Facility) made of aluminium with 4 vertical cylindrical holes (length, 250mm), two for accommodating 10mm Ø capsules and two for 15mm Ø capsules. Length of each capsule: 50mm.

- *HFPIF*

Four tubes of 400mm length, are placed one behind the other in a support, which can be installed on a rail of the pool side facility.
The used aluminium irradiation capsules have an external diameter of 25mm. The length is 80mm or a multiple of 80mm.

- *FIT*

Device for the irradiation of fissile targets for the production of short-lived radionuclides (^{99}Mo). The special targets consist of Al-U sandwich fuel. Up to three cylindrical samples (length 160mm, outer diameter 16mm) can be irradiated in a stainless steel tube (length 480mm, inner diameter 20mm). The device is used in the pool side facility. A similar device has also been constructed for an in-core position (outer sample diameter: 22mm).

- *HIFI*

The device, designed to irradiate Ir-152, is completely fabricated in aluminium.
It uses a maximum of moderator (H_2O) and a minimum of structural material to increase the thermal neutron flux, and consists mainly of five tubes fixed around a central spine.
Each tube can hold five capsules (length 80mm, diameter 10mm or 16mm);

- *PROF*

  Rotating facility in the PSF, reduces by rotating the difference in neutron flux density incident at different parts of the sample. For irradiations polythene bottles with a length of 115 mm and a maximum diameter of 67 mm are used.

- *Conveyor Facility*

  One hydraulic (HR) and two pneumatic conveyor systems (PR 1 and FASY) are available for the radionuclide production in small size capsules in standard shuttles.
  Irradiation times are between a few seconds (FASY) and several days (HR).

16.5.4 Miscellaneous Facilities

- *BONI*

  Capsule for Boron-silicate glass pellet irradiations. Purpose of these experiments is to simulate long-term radiation damage in vitrified fission waste by short-term fission product recoil damage.

- *INTEC*

  Special capsule, in which various types of in-core instruments, intended for application in the TOP experiments (see chapter 16.5.1) are tested under irradiation.

- *TRAMP*

  Rig for determination of the nuclear heating in small samples and for different materials. The rig is used to measure the vertical distribution in various in-core positions of the HFR.

- *TRIP*

  Special design of the TRAMP rig, for use in the pool side facility positions.of the HFR'.

- *TRESON*

  Ellas-type rig to test in-pile ultrasonic thermometers for measuring central temperatures of nuclear fuel.

Fig. 16.8 HFR Petten. View into the reactor hall.

16.6 INFORMATION SHEET No. 6. MAIN CHARACTERISTICS OF HOT CELL FACILITIES

Table 16.2: Main characteristics of hot cell facilities

name of laboratory	year of active operation	scope number of working pos.	active area (m^2)	gamma shielding (material, thickness, density)	leak tightness	special atmosphere	manipulators number and type	remarks
ECN - LSO cells AB, C, D, E	1964	high activity cells 4 in line, 5 working stations	37,5	barytic concrete 120 cm. 3,5 g/cm^3	dynamic	no, unless special in-cell box installed	5 pairs masterslaves heavy duty 1 powermanip. G.E.C. 1 powermanip. Draht & Schrader 1 hoist	cells, AB, C, D, E general purpose use, dismantling, gamma-scanning, dynamic testing
ECN - LSO	1979	metallographic cell-line with head end cell, 10 working stations	10	lead, 11,4g/cm^3 head end cell: 25cm, remaining cells: 18cm	yes	nitrogen or other	tongs in ball-joints and 4 manip. typc CRLG	specially equipped for metallographic examinations and stereo electron microscopy
ECN - LSO G1, G2, G3 and G4 cell	1974	4 medium activity cells, 1 working station each	4 x 1,8	lead and steel, 25 cm. lead ecq.	yes	nitrogen or other	masterslaves, 2 pairs, compact type, CRL model G	G1, G2: cells for chemical work G3, G4: cells for creep measurements
JRC - LSO G5 and G6	1978	2 medium activity cells, 1 working station each.	2 x 1,8	lead and steel, 18cm. lead ecq.	dynamic	no	2 pair CRL	special equipment for dimensional measurements
JRC - LSO H - cell (EUROS)	1983 / 1984 (after reconstr.)	medium activity cell, 2 working stations	ca. 4	lead, 25cm.	a - tight	planned: nitrogen	masterslaves, 2 pairs, compact type, CRL model G	special equipment for remote encapsulation of LMFBR fuel pins into capsules loaded into the HFR
JRC - HFR DMC	1966 (completely overhauled in 1974)	high activity cell 1 working station	ca. 10	cast iron,	dynamic	no	1 pair PYE manip. 1 heavy duty OTER manipulator	direct connection with HFR, cell situated on reactor pool side wing for dismantling of irradiation devices.

16.6.1	*Owner/operator*	*ECN (Netherlands Energy Research Foundation) and the JRC (Joint Research Centre of the Commission of the European Communities)*
16.6.2	*Site*	*Petten, ECN and JRC*
16.6.3	*Associated reactor*	*HFR Petten*

16.7 INFORMATION SHEET No. 7. EQUIPMENT AND TECHNIQUES AVAILABLE FOR POST-IRRADIATION EXAMINATIONS

The combined ECN-JRC facilities can supply all current post-irradiation examination techniques, e.g. dismantling, dimensional measurements, photography, visual inspection, gamma spectrometric scanning, fissiongas sampling and analysis, burn-up determination, metallography and neutrography, mechanical testing comprising tensile (static and dynamic), creep, impact, hardness, electron microscopy, electron microscanning, stereo-scanning, weighing, density determinations;
In addition remote assembly of components, handling of large radiation sources, chemical separations, remote welding.

Limitations:	*shielded flasks*	*- max. weight 25 tons*
	fuel assemblies	*- max. length approx. 2 metres*
	visual inspection	*- periscope 4 - 15x*
	dynamic testing and tensile testing	*- 20 tons*
	creep	*- 3 tons*

16.8 INFORMATION SHEET No. 8. UTILIZATION AND SPECIALIZATION OF THE HOT CELL FACILITIES

16.8.1 *Original purposes of the facilities*

- *Post-irradiation examination of materials irradiated in the HFR, including dismantling of irradiation devices*
- *Support of HFR operation*
- *Support of Dutch nuclear power stations*

16.8.2 *Major modifications subsequently made*

- *reconstruction of cell F1 for metallographic work*
- *construction of a cell (G1) for chemical work (see also Table 16.2)*

16.8.3 *Major modifications planned*

- *modification of material routing system, adaptation of one or two cells and transfer system to α-compatible standards, to meet the requirements set by Pu-enriched fuel.*

- *extension of building, to accommodate additional medium-activity cells for mechanical testing of structural materials.*
- *construction of a medium-activity cell (EUROS) with facility for remote encapsulation of irradiated LMFBR fuel pins into capsules (re-designed ELLAS-, TRIO-, and KAKADU type) to be irradiated in the HFR.*

16.8.4 *Major project and programmes for which the facility presently is used*

See chapter 16.2.4

16.8.5 *Availability of facilities for post-irradiation examination on behalf of other organizations within the European Communities*

Facilities are available.

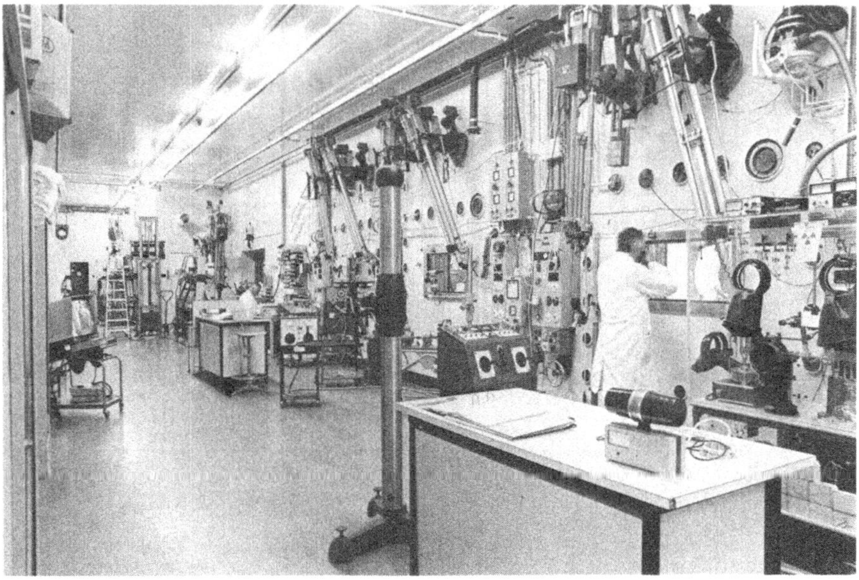

Fig. 16.9 View of some hot cells at the LSO.

17. THE DIDO-REACTOR AT HARWELL, U.K. AND ANCILLARY HOT CELL FACILITIES. INFORMATION SHEETS.

Fig. 17.1 The DIDO reactor with the neutron time of flight tube in the foreground.

17.1 INFORMATION SHEET No. 1. MAIN CHARACTERISTICS OF THE REACTOR.

Name of reactor	DIDO
Organization	UKAEA
Site	Harwell, AERE
In operation since	1956
Power MW (th)	current operating power : 25.5 MW
Moderator	D_2O
Coolant	D_2O - primary; H_2O - secondary.
Fluid and pressure in experiment channels	not applicable

Nominal operation (days/year) 312

Operating cycle (days)
on power 24
shut down 4

Responsible officer (name, address):

DIDO: *K.J. Henry,* *Research Reactors Division*
AERE
Harwell, Didcot, Oxfordshire OX11 ORA
U.K.

17.2 INFORMATION SHEET No. 2 UTILIZATION AND SPECIALIZATION OF THE REACTOR

17.2.1 Original purposes of the reactor

- *Materials and fuels testing in support of civil power reactor programmes.*
- *Neutron Beam experiments for fundamental physics research.*
- *Isotope production.*

17.2.2 Major modifications subsequently made

Progressive increases in power from 10 MW to 25.5 MW, in support of all programmes. Development of special fuel elements to provide more experiment facilities and harder flux spectrum.

17.2.3 Major modifications planned

None

17.2.4 Major projects and programmes for which the reactor presently is used

- *Gas cooled thermal reactor programme.*
- *Fast reactor programme*
- *Water reactor systems*

17.2.5 Loops and irradiation devices specially developed for the reactor

- *Various dynamic gas loops in support of gas-cooled reactor programmes for study of fuels, structural materials, cladding and coolant chemistry.*
- *Irradiation rigs, some of which can be loaded and unloaded at power.*
- *Cold neutron beam source at liquid hydrogen temperature.*
- *High pressure water loops, used with various experiments in reactor core or reflector.*

Main characteristics of specialized irradiation devices are listed in chapter 17.5.

17.3 INFORMATION SHEET No. 3. EXPERIMENTAL FACILITIES OF THE REACTOR

17.3.1 Irradiation positions

Table 17.1: Irradiation positions

type/name	number	active length mm	useful diam. mm	typical unperturbed neutron flux density in $m^{-2}s^{-1}$ thermal	fast $(E > 0{,}1 MeV)$	nuclear heating W/g Al
fuel element						
(i) Mk 5/7	23	600	50	$1{,}8 \cdot 10^{18}$	$0{,}6 \cdot 10^{18}$	4
(ii) S2	2	400	25	$1{,}1 \cdot 10^{18}$	$1{,}6 \cdot 10^{18}$	6
reflector						
2V	9	600	50	$2{,}0 \cdot 10^{18}$	$5 \cdot 10^{16}$	1
4V	5	600	100 *)	$0{,}6 \cdot 10^{18}$	$4 \cdot 10^{15}$	0,2
6V	4	600	150 *)	$2{,}0 \cdot 10^{18}$	$2 \cdot 10^{16}$	1
4VGR	2	600	100	$8 \cdot 10^{16}$	--	--
6VGR	6	600	150	$8 \cdot 10^{16}$	--	--
10VGR	2	600	250	$8 \cdot 10^{16}$	--	--
4H	6	--	100	$1 \cdot 10^{18}$	--	--
6H	1	--	150	$1 \cdot 10^{18}$	--	--
10H	1	--	250	$1 \cdot 10^{18}$	--	--

*) can be subdivided to 50mm or 25mm diameter facilities if required.

17.3.2 Reactivity available for experiments: $7.0\ \%\ \Delta k/k$

17.3.3 Neutron spectra (data): see chapter 17.4

17.3.4 Specialized equipment available for experimenters:
- various handling equipment
- sample changing cells
- data collection facilities

17.3.5 Availability of facilities for irradiation on behalf of other organizations of the European Communities: Facilities are available

17.3.6 Availability of hot cell facilities: Hot cell facilities are available at the Research Reactors Division, and also other large hot cells operated by the Engineering Division and P.I.E. facilities in the Metallurgy Division (see chapter 17.6).

17.3.7 Relevant publications: not communicated

17.4 INFORMATION SHEET No. 4. NEUTRON SPECTRA (DATA)

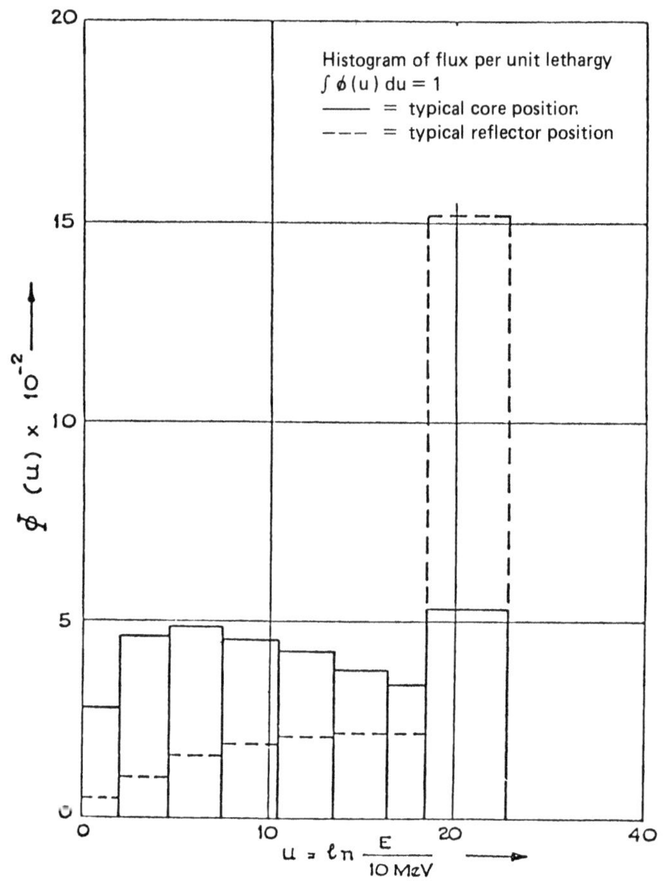

Fig. 17.2 DIDO reactor. Typical neutron spectra.

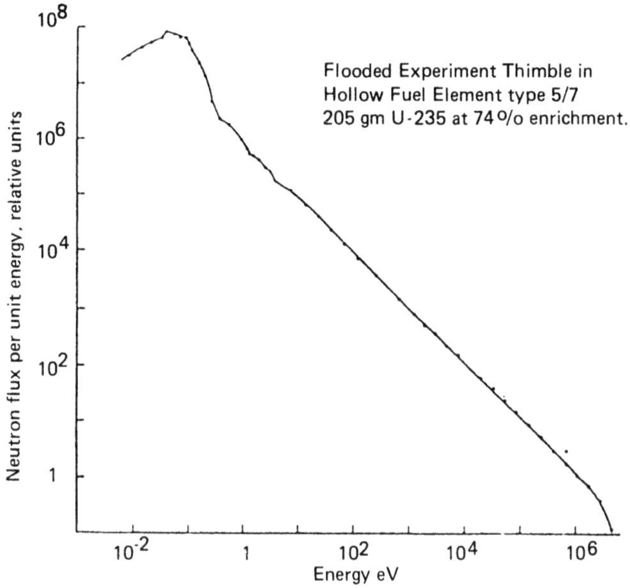

Fig. 17.3

Neutron flux spectra for DIDO and PLUTO reactors.

17.5 *INFORMATION SHEET No. 5. MAIN CHARACTERISTICS OF SPECIALIZED IRRADIATION DEVICES*

- *High Pressure Water Loop*

  A loop constructed in mild steel for the irradiation of fissile specimens in water at the following nominal conditions:

pressure:	*17.2 MPa*
temperature:	*up to 300 °C*
pump differential pressure:	*1 MPa at flow rates up to 2,27 kg/s*
heat rejection capacity:	*3 loops, at 150 kW, 50 kW, 50 kW.*
experiment volumes at high neutron flux:	*600mm long; 78mm Ø*
	600mm long; 38mm Ø (two)
shielding:	*600mm concrete*

- *DIDO Water Loop*

  Loop constructed in stainless steel, for irradiation of fissile specimens in water at the following nominal conditions:

pressure:	*19,3 MPa*
temperature:	*up to boiling point*
flow rate:	*up to 2,5 kg/s*
heat rejection capacity:	*60 kW*
experiment volumes at high neutron flux:	*600mm long; 38mm Ø*

  Special facilities are provided for control of and measurement of water chemistry.

- *Neutron Radiography*

  Neutron radiation facilities for radioactive and non-radioactive specimens and including post-irradiation examination and measurement of specimens within experimental rig assemblies.

17.6 INFORMATION SHEET No. 6. MAIN CHARACTERISTICS OF HOT CELL FACILITIES

Table 17.2: Main characteristics of hot cell facilities

name of laboratory	year of active operation	scope: number of working pos.	scope: active area (m^2)	gamma shielding (material, thickness, density)	leak tightness	special atmosph.	manipulators number and type	remarks
High activity handling building 459	1958	(a) 5	45	1,70m, 2,3 g/cm^3	--	--	20 master slaves NEL model 9	general purpose facility rig & fuel element dismantling, P.I.E. of fuel and reactor components preparation of radio-isotope sources.
	1960	(b) 5	45	1,35m 2,3 g/cm^3	--	--	1 General Mills power manipulation type 350 1 GEC power manip.	
Post Irradiation laboratory	1959	(a) 3	20	1,0 m., 3,2 g/cm^3	--	nitrogen	6 master slave NEL model 9	examination of advanced reactor fuels.
		(b) 9 inter-connected lead cells	14	0,25 m., 11,4 g/cm^3			180 tong manip. Harwell type	
		(c) 20 free-standing lead cells	20	various thickness, 50mm to 250mm 11,4 g/cm^3	to maintain nitrogen atm. of less than 2 °/o O_2			
DIDO handling cell	1957	3	16	1,37 m., 2,3 g/cm^3	--	--	6 type ARGONNE 8 and 9. 1 overhead electrically and hydraulically operated (Limbry & Heron Ltd).	Rig and fuel dismantling or repair
Miscellaneous cells in the Metallurgical and Chemistry Building, equipped for special purposes, e.g. gamma scanning, chemical analysis, etc.								
Miscellaneous cells in DIDO & PLUTO reactors for rig sample changing, absorber maintenance, etc.			up to 1	0,25 m., 11,4 g/cm^3	--	--	up to 5 tong manip. Harwell type	

17.6.1 Owner/operator

High activity handling	*owner and operator : Engineering Division*
Post irradiation lab.	*owner: Metallurgy Division; operator: Engineering Div.*

17.6.2 Site — *Harwell, AERE*

17.6.3 Associated reactor — *primarily DIDO and PLUTO (Harwell) and PFR (Dounreay), but material is also received from reactors in other countries.*

17.7 INFORMATION SHEET No. 7. EQUIPMENT AND TECHNIQUES AVAILABLE FOR POST-IRRADIATION EXAMINATION

- *Equipment for dismantling of large irradiation rigs and experimental fuel assemblies.*
- *Equipment for the detailed examination of fuels and components, including*
  - *measurement of length and diameter changes utilising electrical transducers.*
  - *collection of fission product gases using laser piercing techniques.*
  - *preparation of materials for examination on optical, electron and stereo-scanning microscopes.*
  - *precision gamma scanning of fuels.*
  - *machining of irradiated materials in mechanical test pieces.*
  - *equipment for mechanical properties testing - tensile creep, fatigue, etc.*
  - *re-encapsulation of irradiated materials.*
  - *preparation and encapsulation of radio-isotopes, cobalt, caesium, californium.*

Relevant publications

- *J.E. Bown, et al.*
  The High-Activity Handling Building at A.E.R.E. Harwell.
  Proceedings 2nd United Nations Conf. on the Peaceful Uses of Atomic Energy, Vol. 17, pp. 631 - 644 (Sept. 1958).
- *J.E. Bown, A.B. Ritchie*
  Operations in the High Activity Handling Building B.459, A.E.R.E, Harwell.
  Proc. 8th Conf. Hot Labs. Equip., TID 7599, 1, 50 USAEC Div. Tech. Info. (1960).
- *H.J. Venables*
  Remote Handling of Irradiated Alpha-Active Materials in the Metallurgy Division, A.E.R.E., Harwell.
  Proc. 8th Conf. Hot Labs. Equip., TID 7599 1, 78, USAEC Div. Tech. Info (1960).
- *A.B. Ritchie, G.A. Serrels*
  Dismantling and Machining Operations on High Activity Rigs at Harwell.
  Proc. Symp. High Activity Hot Lab. Working Methods, Grenoble 1, 329 (1965).
- *A.B. Ritchie*
  Safety Experience and the Control of Hazards in Hot Cell Operations at Harwell
  Proc. 15th Conf. Remote Systems Technology p. 167 (1967).
- *D. Campbell*
  Post Irradiation Examination Facilities and Techniques
  Journal of the Inst. of Nucl. Engs., Vol. 11, no. 5, p. 136 (1970).

- *J.B. Rich et al.*
  Techniques employed in the examination of irradiated prototype fuel pins for high temperature gas-cooled reactors.
  Proc. Symp. Post-Irradiation Examination Techniques.
- *P.E. Brown, et al.*
  Lead-shielded Micro-handling Cell for Coated Fuel Particles.
  A.E.R.E. R.6702.

17.8 *INFORMATION SHEET No. 8. UTILIZATION AND SPECIALIZATION OF THE HOT CELL FACILITIES.*

17.8.1 *Original purposes of the facility*

The two major post-irradiation examination facilities were designed in support of a materials testing programme for the British Nuclear Power Programme. This is still the main purpose of the facility.

17.8.2 *Major modifications subsequently made*

No major modifications have been made to the concrete cell facilities, but the equipment within is frequently changed, dependent on the research programme.
In the lead cell facility a number of the cells were regrouped and linked together to achieve better productivity.

17.8.3 *Major modifications planned*

A new cell including the provision of a new optical micrsocope is being built.

17.8.4 *Major projects and programmes for which the facility currently is used.*

- *Fact reactor project*
- *Advanced gas-cooled reactor project.*
- *water reactor system.*

17.8.5 *Availability of facilities for post-irradiation examinations on behalf of other organizations within the European Community.*

Facilities are available.

18. THE PLUTO REACTOR AT HARWELL, U.K. AND ANCILLARY HOT CELL FACILITIES. INFORMATION SHEETS.

Fig. 18.1 The PLUTO reactor building.

18.1 INFORMATION SHEET No. 1. MAIN CHARACTERISTICS OF THE REACTOR.

Name of reactor	*PLUTO*
Organization	*UKAEA*
Site	*Harwell, AERE*
In operation since	*1957*
Power MW (th)	*current operating power : 25,5 MW*

Moderator	D_2O
Coolant	D_2O - *primary;* H_2O - *secondary*
Fluid and pressure in experiment channels	*not applicable*
Nominal operation (days/year)	*312*
Operating cycle (days)	
on power	*24*
shut down	*4*

Responsible officer (name, address):

PLUTO : *K.J. Henry* *Research Reactors Division*
AERE
Harwell, Didcot, Oxfordshire, OX11 ORA
U.K.

18.2 INFORMATION SHEET No. 2 UTILIZATION AND SPECIALIZATION OF THE REACTOR

18.2.1 Original purposes of the reactor

- *Materials and fuels testing in support of civil power reactor programmes.*
- *Neutron Beam experiments for fundamental physics research.*
- *Isotope Production.*

18.2.2 Major modifications subsequently made

- *Progressive increases in power from 10 MW to 25,5 MW in support of all programmes.*
- *Development of special fuel elements to provide more experiment facilities and harder flux spectrum.*

18.2.3 Major modifications planned

- *Replacement of reactor top shield plug, to reposition and improve vertical experiment facilities.*

18.2.4 Major projects and programmes for which the reactor presently is used

- *Gas cooled thermal reactor programme.*
- *Fast reactor programme.*

18.2.5 Loops and irradiation devices specially developed for the reactor

- *Various dynamic gas loops in support of gas-cooled reactor programmes for study of fuels, structure materials, cladding and coolant chemistry. Also effect of power cycling on fuels and cladding materials.*
- *Irradiation rigs, some of which can be loaded and unloaded at power.*

Main characteristics of specialized irradiation devices are listed in chapter 18.5.

Fig. 18.2 PLUTO Reactor. Top reactor shield with some irradiation rig handling flasks.

18.3 INFORMATION SHEET No. 3. EXPERIMENTAL FACILITIES OF THE REACTOR

18.3.1 Irradiation positions

Table 18.1 : Irradiation positions

type/name	number	active length mm	useful diam. mm	typical unperturbed neutron flux density in $m^{-2}s^{-1}$ thermal	fast (E > 0,1 MeV)	nuclear heating W/g Al
fuel element						
(i) MK 5/7	24	600	50	$1,8 \cdot 10^{18}$	$6 \cdot 10^{17}$	4
(ii) S2	2	400	25	$1,1 \cdot 10^{18}$	$1,6 \cdot 10^{18}$	6
reflector						
C.C.A.	3	600	*)	$1,6 \cdot 10^{18}$	10^{16}	0,5
7V	4	500	175 *)	$1,5 \cdot 10^{18}$	$5 \cdot 10^{15}$	0,3
4V	4	200	100 *)	$5 \cdot 10^{17}$	$2 \cdot 10^{15}$	0,1
4VGR	6	200	600 *)	$8 \cdot 10^{16}$	$7 \cdot 10^{13}$	neg.

*) can be subdivided to 50mm or 25mm diameter facilities.

18.3.2 Reactivity available for experiments: 7,0 °/o Δk/k

18.3.3 Neutron spectra (data): see chapter 18.4

18.3.4 Specialized equipment available for experimenters:
- various handling equipment
- sample changing cells
- data collection facilities
  etc.

18.3.5 Availability of facilities for irradiation on behalf of other organizations of the European Communities: Facilities are available

18.3.6 Availability of hot cell facilities: Facilities are available at the Research Reactors Division. Also other large hot cells, operated by Engineering Division, and P.I.E. facilities in the Metallurgy Division (see chapter 17.6).

18.4 INFORMATION SHEET No. 4. NEUTRON SPECTRA (DATA)

See also Fig. 17.3

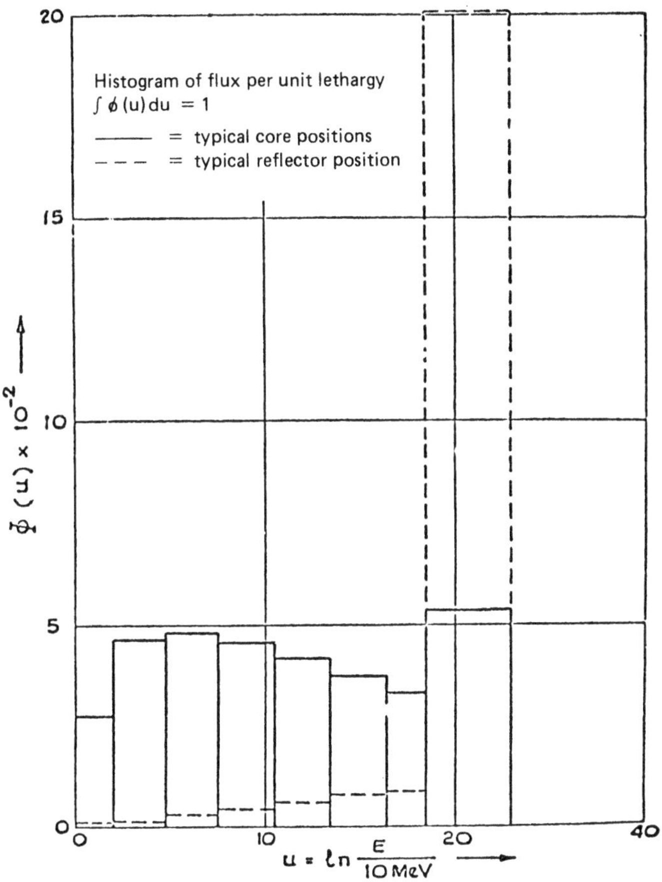

Fig. 18.3 PLUTO Reactor. Typical Neutron Spectra.

18.5 INFORMATION SHEET No. 5. MAIN CHARACTERISTICS OF SPECIALIZED IRRADIATION DEVICES.

- *Gas Loops*

  A number of dynamic gas loops have been operated in which helium or other coolant gases at pressures up to 7 MPa and gas temperatures up to 400 ^{o}C are circulated over fissile fuel specimens. Fine control of coolant gases down to 1 v.p.m. total impurities is possible and controlled additions of moisture or other impurities is provided for.
  Automatic control of gas flow rates and specimen temperatures is provided. Neutron flux can be finely controlled using He^3 gas as an absorber and power cycling with varying ratios (up to approx. 5) with pre-determined ratio change characteristics are achieved using the same gas.

18.6 INFORMATION SHEET No. 6. MAIN CHARACTERISTICS OF HOT CELL FACILITIES

See chapter 17.6

18.7 INFORMATION SHEET No. 7. EQUIPMENT AND TECHNIQUES AVAILABLE FOR POST IRRADIATION EXAMINATION

See chapter 17.7

18.8 INFORMATION SHEET No. 8. UTILIZATION AND SPECIALIZATION OF THE HOT CELL FACILITIES

See chapter 17.8

19. HOT CELL FACILITIES AT CASACCIA, ITALY.

Fig. 19.1 Front view of the high activity cell line.

19.1 INFORMATION SHEET No. 1. GENERAL INFORMATION ON THE HOT CELL FACILITIES

Name	*OPEC Hot Lab*
Organization	*CNEN - Italy*
Site	*CNEN - CSN Casaccia, (Roma)*

Responsible officer (name, address):

Dr. G. Pugnetti	*CNEN* *CSN della Casaccia* *C.P. 24000* *I - 00100 Roma A.D.*

19.2 INFORMATION SHEET No. 2. MAIN CHARACTERISTICS OF HOT CELL FACILITIES

Table 19.1 : Main characteristics of hot cell facilities

name of laboratory	year of active oper.	scope		gamma shielding (material, thickness, density)	leak tightn.	special atmosph.	manipulators number, type (incl. spare manipulators)
		active area (m^2)	number of working positions				
OPEC 1	1961	24	5	concrete, 0,75 m, 3,5 g/cm^3	only γ	no	10 MOD 8 1 NEL 100
OPEC 2	non nuclear tests during 1980	60	10	concrete, 1 m, 3,6 g/cm^3 or 0,65 m, 3,6 g/m^3	α - γ	provision for inert atmosph.	10 MOD 8 (HD) 2 ABC (Heavy mechanic arms) 10 MOD G

19.2.1 *Owner/operator :* CNEN, Casaccia, Roma

19.2.2 *Associated reactor :* None

19.2.3 *Relevant publications :*

- Cacciari, Cesarano, Lepscky, Nepi : Hot facilities of CNRN. Proc. 8th Conference Hot Laboratory and Equipment 167 (1960).
- Cesarano, Cudia, Evangelisti, Lauro, Vescia : The Alpha-Gamma Radiometallurgy Laboratory at Casaccia Nuclear Center - Design and Construction. Proc. 19th Conference on Remote Systems Technology 145 (1971).
- C. Cesarano et al. : Design and safety criteria of the α/γ hot laboratory of CNEN, Italy. IAEA, SM. 209/29/1976.

Fig. 19.2 Internal view of the non-destructive test cell.

19.3 INFORMATION SHEET No. 3. EQUIPMENT AND TECHNIQUES AVAILABLE FOR POST-IRRADIATION EXAMINATIONS.

19.2.3 Equipment and techniques available :

- *Non destructive tests (visual, gamma scan, gamma densitometry, metrology, X-ray radiography)*
- *Piercing for fission gas collection and analysis*
- *Dismantling of fuel and capsules*
- *Macro and Microscopy*
- *Radial sampling and radiochemical analysis*
- *Annealing; HD determination; O/Me determination*

19.4 INFORMATION SHEET No. 4 UTILIZATION AND SPECIALIZATION OF THE FACILITY

19.4.1 Original purposes of the OPEC 1-facility

- *Any kind of experiments requiring hot cell facilities*

19.4.2 Major modifications subsequently made

- *Enlarging the office and low activity laboratory space in order to afford a comprehensive programme of post-irradiation analysis of ceramic nuclear fuels.*
- *Building a new wing (OPEC 2) for α-γ post-irradiation work on Pu fuels.*

19.4.3 Major modifications planned

- *Ventilation system*

19.4.4 Major projects and programmes for which the facilities presently are used.

- *CIRENE project*
- *Basic research and development*
- *Miscellaneous requirement of other programmes*

19.4.5 Availability of facilities for post-irradiation examinations on behalf of other organizations within the European Community:

Facilities are not available at the moment.

ANNEX TABLE OF MATERIALS TESTING REACTORS IN THE COUNTRIES OF THE EUROPEAN COMMUNITY

Country	Name of reactor	Owner	Site	In operation since	Thermal power MW	Neutron flux density in 10^{18} $m^{-1}s^{-1}$ Thermal	Fast (E > 0,1 MeV)	Hot cell facilities
BELGIUM	BR 2	CEN/SCK	Mol	1961	80 - 125	9	7	ATHA, LHMA
DENMARK	DR 3	Risø National Laboratory	Risø	1960	10	1,5	0,45	Hot cells
F.R. GERMANY	FR 2	KfK	Karlsruhe	1961	44	0,95	0,06	Dismantling cell FR 2 and also for KNK II: Hot cells building 701 and TPL
	KNK II	KfK	Karlsruhe	1977 *)	58 (20 MWel)	--	20 **)	
	FRJ 1 (MERLIN)	KFA	Jülich	1962	10	1	0,4	Hot cells (HZ and BZ) and "Chemiezellen"
	FRJ 2 (DIDO)	KFA	Jülich	1962	23	2	1	
	FRG 2	GKSS	Geesthacht	1963	15	1,5	1,2	Hot cells
	FRG 1	GKSS	Geesthacht	1958	5	0,5	0,4	
FRANCE	OSIRIS	CEA	Saclay	1966	70	3	5	LECI, CELIMENE, SEMIRAMIS
	ORPHEE	CEA	Saclay	1980	14	3	--	
	SILOE	CEA	Grenoble	1963	35	5	5	Siloe Hot Cell, } LAMA/DMG
	MELUSINE	CEA	Grenoble	1958	8	0,9	1	Melus. Hot Cell } LAMA/DMG
	HFR	ILL	Grenoble	1971	57	8 - 13	--	LTHA
	TRITON	CEA	Fontenay-aux-Roses	1959	6,5	0,6	0,35	Triton Hot Cell Hot Cell Facilities FAR
ITALY	ESSOR	EEC	Ispra	1968	25 (18)	4	1,1	ADECO, ATFI, LMA.
NETHERLANDS	HFR	EEC	Petten	1961	45	3,3	4,5	DM Cell, LSO ***)
UNITED KINGDOM	DIDO	UKAEA	Harwell	1956	25,5	2,0	1,6	High activity handling, bldg. 459, P.I. Lab.
	PLUTO	UKAEA	Harwell	1957	25,5	1,8	1,6	DIDO handling cell

*) with fast core **) E > 10 keV ***) owned and operated by the Netherlands Energy Research Foundation (ECN)

GPSR Compliance

The European Union's (EU) General Product Safety Regulation (GPSR) is a set of rules that requires consumer products to be safe and our obligations to ensure this.

If you have any concerns about our products, you can contact us on ProductSafety@springernature.com

In case Publisher is established outside the EU, the EU authorized representative is:

Springer Nature Customer Service Center GmbH
Europaplatz 3
69115 Heidelberg, Germany

Batch number: 10584448

Printed by Printforce, the Netherlands